Energy

Selection Guide Series Titles

Energy

A MULTIMEDIA GUIDE
FOR CHILDREN AND YOUNG ADULTS

Judith H. Higgins

Neal/Schuman Publishers   ABC-Clio, Inc.

New York   Santa Barbara   Oxford

This book is for Ted Higgins and Ted Galloway

Published by the American Bibliographical Center—Clio Press
2040 Alameda Padre Serra
Box 4397
Santa Barbara, CA 93103
In association with Neal-Schuman Publishers, Inc.

Printed and bound in the United States of America

Library of Congress Cataloging in Publication Data

Higgins, Judith, 1930–
    Energy.

    (Selection guide series: no. 2)
    Bibliography: p.
    Includes index.
    1. Power resources—Bibliography. 2. Power
resources—Information services. I. Title.
II. Series.
Z5853.P83H53   [TJ163.2]   016.3337   78-15611
ISBN 0-87436-266-0

Table of Contents

Foreword

Energy—it is a subject of immense importance to all of us. Understanding it is the first step towards the assurance that we will have sufficient energy to meet our future needs.

The Department of Energy was not conceived with the idea that it would be the sole instrument designed to "solve" the national energy problem. This must be done in concert with individual citizens working collectively towards a common end.

The purposes of the Department are to stimulate research and development, encourage conservation, assist with commercialization of alternate energy sources, and provide incentives to help society bridge the gaps between supply and demand. The Department has many goals to achieve, but the central theme in working to achieve them is that it will be the actions of the American public that determine the energy future of the Nation. The energy picture is a complex one, partly because we have taken energy for granted for so many decades, but essentially because energy cuts so deeply into the very heart of modern civilization.

While the Department does not endorse specific publications, it does encourage the flow of information. We recognize that before a problem can be solved, it must be understood. The materials listed in this guide can be important sources for developing understanding of energy issues, for taking that first step towards a solution.

James Bishop, Jr.
Director
Office of Public Affairs
Department of Energy

Preface

One of the few points upon which energy analysts agree is that the United States, with six percent of the world's population, consumes at least one-third of the world's energy. And the vast bulk of this energy still comes from finite sources—oil, natural gas, and coal. The recent energy crisis makes it clear that energy is a complex subject and that a national energy plan is essential.

What is not clear is how the energy needs of the future should be met. Many of the decisions regarding energy development, conservation, and use will be made by those students who are sitting now in American classrooms. These will be not simply scientific decisions but—as our relations with the Organization of Petroleum Exporting Countries (OPEC) and the clashes between environmentalists and energy industry expansionists have shown—political, economic, and societal decisions as well. Where will we put our money and our time, and with what results?

Energy: A Multimedia Guide for Children and Young Adults attempts to provide a source of energy information for American teachers; librarians; and media, curriculum, and information specialists. It concentrates on material or sources of material that—in the opinion of the editor and of qualified teachers at the elementary and secondary levels—should be useful with students in kindergarten through twelfth grade. It acknowledges that the ability range in these grades is enormous, and that ideally there should be material for every level of understanding.

Every effort was made to make this guide an inclusive collection of quality energy materials. Only in-print materials are listed. My goal has been to save the time and effort of the practicing librarian, teacher, or curriculum designer by evaluating or describing a wide range of materials from governmental, commercial, educational, industrial, and professional sources. These are sources which should be useful in the acquisition of energy materials for a media collection or for aid in designing or amending curriculum. The works of small, perhaps obscure publishers and producers are included as well as those of large, familiar ones, for this is an area with many specialists and specialized sources. Some of the most interesting materials have come from small firms concentrating on a local problem or explaining a very specific idea.

A word of caution: this is a highly partisan field, one peppered with conflicting statistics and confusing jargon. While many of the materials evaluated or described here are produced by firms which undoubtedly have attempted to present a fair picture of their subject, some can only be considered propaganda. Too often students take what they are given as gospel, and too often material is presented that way. It is our responsibility as educators to balance the picture: not only to make available a variety of materials but to explain why these materials say what they do. If our goal is to help students make intelligent observations and decisions, we must provide some background as well as foreground.

The materials in this guide are presented in four broad sections: 1. Multimedia Materials, K–12, 2. Curriculum Materials, 3. Selection Aids, and 4. Further Sources of Information. Author, Title, and Subject Indices as well as a Directory of Publishers and Distributors are included. A more lengthy discussion of the content, arrangement, and criteria for inclusion of each of these sections follows.

Multimedia Materials

Within this section annotated energy titles are arranged by specific subjects. These are: General, Coal, Electric Power, Energy Conservation, Energy–Environmental Aspects, Energy–Political and Economic Aspects, Engines, Gas, Geothermal Power, Heat Pump, Methane, Methanol, Nuclear Energy, Oil, Pedal Power, Reference, Solar Energy, Water Power, Wind Power, Wood. Titles are further broken down into print (books, pamphlets, booklets, monographs, newsletter issues) and nonprint (films, filmstrips, videotapes, slides, audiotapes, charts, posters, kits). Within each of these categories, the items are listed in ascending grade level. Grade 12 is not necessarily the final audience for material so keyed, but it is the top grade within the concern of this book.

Inevitably, the arrangement is somewhat arbitrary. Many of the materials listed overlap subject categories. For a finer examination of the contents, the subject index of this book should be consulted.

In some subject categories there is a wealth of materials; in others, very little. The number of titles reflects both the existing market (some topics, such as solar energy, apparently have caught the public and publishing eye more than others) and the editor's reluctance to include dated material which has been reissued in new packages to capitalize on the interest in energy.

More than 1000 publishers, producers, distributors, utility companies, and professional, educational, industrial, and environmental organizations were contacted during the fall and winter of 1977-78. They were asked to supply evaluation copies of materials currently in print that would fall within the scope of this guide. The tools used as a starting point to locate potential sources of energy information included the following: *Subject Guide to Books in Print,* the *Subject Guide to Children's Books in Print, Audiovisual Market Place: A Multimedia Guide,* the *Children's Catalog, The Core Media Collection for Secondary Schools,* the *AAAS Science Film Catalog, Educational Media Yearbook, The Elementary School Library Collection: A Guide to Books and Other Media,* and the article "An Energy Source Directory" in the January 1, 1978, issue of *Library Journal.* Other sources were contacted as the editor became aware of them. Reference and science-technology librarians in the White Plains and Yonkers (New York) public libraries offered suggestions.

The criteria used for inclusion were: currency, clarity and completeness of presentation, quality of format, and relevance of the material to the elementary and secondary curriculum. Materials felt to be too technical or too advanced for even the bright secondary student were examined and rejected. Some consumer-oriented materials are included since it was felt they could be well utilized in consumer education, industrial arts, and occupational education programs. Textbooks generally were not considered. I drew upon the expertise of science, social studies, and industrial arts teachers at the Valhalla (New York) junior-senior high school as well as that of instructors in the occupational educational program of the Mid-Westchester Center for Occupational Education of the Southern Westchester Board of Cooperative Educational Services (BOCES). In evaluating how-to books we have tried to include only those that utilized materials that would be readily available to teachers and students in occupational education programs and that were not beyond the technical ability of most students in these programs.

Several "source" books are listed and evaluated in this Multimedia Materials section rather than in the section on Curriculum Materials because they were found in reality to be background readers on energy rather than curriculum outlines. Certain subject areas were particularly difficult to evaluate. The introduction to Sheldon Novick's *The Electric War* (San Francisco, Calif.: Sierra Club Books, 1976) sums up the problem, for example, posed by materials on nuclear power: "Much of what is said publicly about nuclear power is untrue; much else is secret. Accounts of the controversy concerning it seem confusing and are often contradictory."

Curriculum Materials

This section includes curriculum guides, courses, mini-units, manuals, modules, and activity packages. Titles are arranged by ascending grade level. Many of the items listed have parts geared to various grade levels, and they are arranged here according to the unit covering the lowest grades. For example, a title with an overall grade span of K-12 but with a unit for grades K-3 will precede a title also spanning grades K-12 but with a unit for grades K-6. Items are described rather than evaluated. If there is any limitation on their availability, e.g., they are free to educators in one state but not to out-of-state persons, it is noted.

Selection Aids

Periodicals that regularly review science, social studies, or general materials on energy subjects are described and their reviewing policies explained. They are arranged alphabetically by title. Bibliographies that cover both print and nonprint materials are included and are arranged alphabetically by author's name. An effort was made to eliminate lists that referred to materials no longer available.

Further Sources of Information

This section is subdivided as follows:

Periodical Indices. Periodicals often provide the bulk of our current information. The indices that, in the opinion of the editor and experienced reference librarians, would be most useful in obtaining energy information from periodicals are given here. A list of subject headings culled from these indices is provided to save the user valuable searching time.

U.S. Government Sources. Since much of our energy material comes from the federal government at a reasonable price, every effort was made to bring the sources of federal-government material up to date. The help of the U.S. Government Printing Office was invaluable in this regard.

Free-Loan and Rental Films. Free-loan films from many of the major energy companies are noted in the Multimedia Materials, Selection Aids, U.S. Government Sources, and Industrial, Citizen, Professional, and Public Sources sections. Additional sources of free-loan films and videotapes are given here. Although the source for renting films evaluated in the Multimedia Materials section is noted with each entry, many schools and libraries prefer to use university film-rental libraries to avoid multiple orders. A mailing was done to all university film-rental libraries that are listed in the Educational Film Library Association publication *University and College Film Collections: A Directory,* 3rd. ed. (New York: EFLA, 1974-75), that had a total of more than 100 titles in their general collection, and that did not limit their film service to a particular group. Respondents who indicated that their collections now include films on energy are listed here. Geographical restrictions on rentals are noted.

Industrial, Citizen, Professional, and Public Sources. These organizations not only have current titles, which are pointed out in the annotations here, but are likely continuing sources of information on energy.

Data Bases. As the computer has entered both the school and the public library, more and more of today's students have access to data bases, which formerly were the province of the college or graduate student, and they are familiar with searching techniques.

Source Directory and Indices

The Directory of Publishers and Distributors gives ordering addresses for all the titles annotated in the numbered sections of the book.

The final pages provide three indices: author, title, and subject. Both the author and title indices provide access to the numbered entries which make up the Multimedia Materials section, the Curriculum Materials section, and the Selection Aids section. The subject index provides specific access to the contents of both the unnumbered Further Sources of Information section, and to numbered entries.

Acknowledgments

No book is ever produced alone. For their advice and assistance I am indebted to Bob Beitscher, Delores Booker, Joyce Bush, Dominick Canestra, Emily Carl, Marge Castro, Muriel Clarke, Dr. Al Cohen, Terry Cotter, Clark Darling, Paul DeCelie, Steve DeMicco, Frank D. Doble Jr., Bette Fraim, Bob Gaglione, Berry Gargal, Bill Handley, Carol Hart, Grace Holden, Ted Jednak, John Kager and the media staff of the Westchester Community College Library, Bill Katz, Diana Koster, Ralph Lagani, Sara Lalli, Pauline Maluski, Iris Masback, Robert Maus, Florence Modell, Hugh McKenna, Charlotte Newman, Millie Poncé, Rose Proietti, Dr. Don H. Richardson, JoAnn V. Rogers, Bob Schneider, Blanche Seaman, Sharon Siegler, Lada Simek, Nancy Storck, Ferd Ventare, Bob Williams, Betty Zeil, and above all, to Bob Obuck, whose knowledge and jibes spurred the book's completion. I am grateful as well to the many producers and publishers who offered their materials for evaluation and to Maureen Crowley and Pat Schuman, who saw the manuscript through.

Judith Higgins
August, 1978

Part One

Multimedia Materials, K-12

General

PRINT

1. Pine, Tillie S., and Levine, Joseph. **Energy All Around.** Illus. by Joel Schick. New York: McGraw-Hill, 1975. 48pp. $5.72 plb. Gr. 2–4.

Using comical illustrations, this beginning science book gets across some basic ideas about what energy is, how energy changes, and how we use energy in the form of food and fossil fuels. It gets progressively more difficult, working up to terms like turbines, generators, and plutonium. Younger children will need help with it.

2. Israel, Elaine. **The Great Energy Search.** New York: Messner, 1974. 64pp. $4.29 plb. Gr. 3–6.

A once-over-lightly, heavily illustrated introduction to our energy problems, the fuels we use, and where our future energy sources may lie. Although the information on the lack of an energy policy is dated, the questions this title raises about energy choices and conservation are well handled.

3. Adler, Irving. **Energy.** New York: Day, 1970. 48pp. $5.79 plb. Gr. 4–6.

This title defines energy, the forms it may take, where it comes from, and how it's used. There is a good explanation of the changes in forms of energy, e.g. at a power plant, where the chemical energy of fuel is changed into heat energy, then into kinetic energy, and finally into electrical energy. Three thoughtful closing pages describe the energy imbalance in the world and what might be done to rectify it. Illustrations are clear and the pronunciation of unfamiliar words is given in parentheses.

4. Gulf Oil Corporation. **An Eagle Eye on Energy.** Houston, Tex.: Gulf Oil, n.d. 16pp. free. Gr. 4–6.

In this cartoon-pamphlet an American eagle explains what energy is and what the energy crisis is all about. It describes the three fossil fuels which are the United States's primary sources of energy, as well as other sources of energy. It also suggests several ways to conserve energy at home. A chart, *Energy: Where It Comes From, Where It Goes*, accompanies the pamphlet.

5. Smith, Norman F. **Energy and Environment.** Austin, Tex.: Steck-Vaughn, 1974. 48pp. $3.19 plb. Gr. 4–6.

A number of energy topics including early use of energy, fossil fuels, energy and transportation, electrical energy, and new sources of energy are briefly and clearly discussed in this book. The contents of each section are summarized in introductory statements. The closing section, "How Much Energy Do We Really Need?," is thought provoking; all but one of the five suggested activities make sense. Although the format is not particularly attractive and the book lacks an index, topics can be picked from it easily, and its comprehensive coverage is appropriate for the intended age group.

6. Doty, Roy. **Where Are You Going with That Energy?** Chicago Museum of Science and Industry Series. New York: Doubleday, 1977. 62pp. $5.95. Gr. 5–9.

This lively introductory book is brightened by Doty's two-page drawings. Short, self-contained blocks take up fire, water, wind, animal power, coal, steam, animal oils, petroleum, electricity, natural gas, explosives, the internal combustion engine, high-energy fuels, energy from the atom (not including fusion), solar energy, geothermal energy, energy from the ocean, oil shale and sands, fuel cells and batteries, and the direct conversion of solar energy into electricity. The book concludes with a section on the worldwide nature of the energy problem. The author emphasizes that much still is not known about energy and energy sources. A bolt of lightning runs through the book to remind readers of man's constant search for new sources of energy.

7. Woodburn, John H. **Energy.** Merit Badge Pamphlet No. 3335. North Brunswick, N.J.: Boy Scouts of America, 1978. 64pp. $.55 pap. Gr. 6 up.

Although intended as a guide for Boy Scouts and Explorer Scouts working toward the Energy Merit Badge, this pamphlet also contains much information of general interest. It provides a basic introduction to the subject of energy and outlines the eight requirements for this merit badge. To help readers meet these requirements, the pamphlet provides information on building a solar energy box, what energy is and what it does, sources and supplies of energy, the conversion of energy in engines, and energy use and conservation. A table showing the average annual electrical energy consumption by appliances and a "home energy check" are included, as are a list of books recommended by the American Library Association's Committee to Scouting and a list of free or low-cost publications relating to energy.

8. Michelsohn, David Reuben, and the Editors of Science Book Associates. **The Oceans in Tomorrow's World: How Can We Use and Protect Them?** New York: Messner, 1972. 189pp. $5.29 plb. Gr. 6–9.

Chapter six, "Power from the Sea," provides one of the few explanations of tidal power as a source of energy and briefly discusses ocean-temperature differentials and deuterium as potential energy sources. Portions of chapters one and four discuss oil pollution and offshore oil drilling.

9. Rothman, Milton A. **Energy and the Future.** New York: Watts, 1975. 114pp. $5.90 plb. Gr. 6–10.

In this brief, state-of-the-art survey, the author describes what energy is, where it comes from, how it is generated, and how it contributes to our lives. He raises valid questions about the price we pay for producing and consuming each form of energy. Offering no easy solutions, Rothman believes that "the one basic attitude that must change is the idea that *more and more* is the same as *better and better.*"

10. Halacy, D. S., Jr. **Earth, Water, Wind, and Sun: Our Energy Alternatives.** New York: Harper, 1977. 186pp. $8.95. Gr. 6–12.

Halacy, an experienced science writer, debunks a nuclear future and takes a generally realistic look at the state of the art of energy, including geothermal springs, hydropower, tides, sea thermal energy, wind energy, solar energy, and biofuels such as wood, methane, sewage gases, trash and garbage. Well written, with small but not unreadable type, this title also provides a history of the attempts to make each energy source work.

11. American Public Power Association. **APPA Consumer Folders.** Washington, D.C.: American Public Power Assn., n.d. 6pp. $9 per 100; minimum order 500 of each title. Gr. 7–12.

Designed for utility companies to distribute to customers, this series of small pamphlets provides basic information on certain energy topics from the electric utility point of view. They may be ordered directly in quantity or obtained from the local electric utility. Useful titles for science, consumer education, or home economics classes are: "How Insulation Can Save You Money," "Six Ways to Save Money on Your Utility Bills," "How to Get the Most from Your Appliances," "Who Owns Your Electric Utility?," "Electricity and Your Environment," "How Your Electric Utility Operates," "How Much Electricity Do Your Home Appliances Use?," "How to Get the Most from Your Electric Clothes Dryer," "How Your Electric Meter Works," "How to Save Money by Heating Your Home Efficiently," "Everything You Always Wanted to Know about Energy Conservation," "Heat Pump: The Energy Conservation Machine," "The Struggle to Hold the Line on Electric Rates."

12. Burbank, James, and Pearson, Eloise V. **Energy Sources: A New Beginning.** Boulder, Colo.: Univ. of Colorado, Educational Media Center, n.d. 9 study guides, 1 teacher's guide. $1.75 ea. Gr. 7–12.

These study guides were designed to accompany each of the nine films in the Energy Sources: A New Beginning series. Each guide contains a glossary of terms, describes how the film and the guide are to be used, and examines the history, present technology, and future prospects for a specific energy resource. A summary, class projects, study questions, and a short bibliography round out each guide. Although they are best used to supplement the film series, these guides could also stand alone as they include a great deal of solid information. The guide to *Nuclear Energy*, for example, traces the history of atomic energy development as far back as 1800, and discusses the hazards of plutonium and the possibilities for fusion power. The teacher's guide, *Teaching the Energy Series*, raises ten critical questions about energy, describes each module, and suggests how the modules can be used. It also includes a 21-page annotated bibliography of related materials.

13. Clark, Peter, and Landfield, Judy. **Natural Energy Workbook Number Two.** Berkeley, Calif.: Visual Purple, 1976. 128pp. $3.95 pap. Gr. 7–12.

Billed as a "comprehensive text on the application of locally regenerative resources (sun, wind, water and photosynthetic fuels) to local energy needs," this book can also serve as an inexpensive teacher's handbook and resource for classroom or outside projects related to energy. Printed on newsprint with wide margins to encourage note-making, it is packed with information and illustrations. It could also be used as a supplemental text for science classes.

14. Fowler, John W. **Energy Environment Source Book.** Washington, D.C.: National Science Teachers Assn., 1975. 279pp. $4 pap. Gr. 7-12.

This objective, highly readable source book was written for teachers who want to incorporate material on energy into their classes. It is divided into two parts. Volume one deals with the economic, political and social issues involved in the energy-environment debate. Volume two provides the scientific and technical background needed for a thorough understanding of the issues outlined in the first volume. Technical appendices provide a glossary; a mathematics primer; a section explaining the relationship between force, work energy, and power; a section on heat and heat engines; and sections on the generation and transmission of electricity, and on nuclear energy. The author's intention is to take teachers of science, social studies, and the humanities beyond their own disciplines, and he encourages teachers to select material from both sections of the book.

15. Halacy, D. S., Jr., **The Energy Trap.** New York: Four Winds, 1975. 143pp. $6.95. Gr. 7-12.

The title refers to Halacy's contention that "we have trapped ourselves in the energy revolution" as a result of our desire for continuing growth and comfort. In self-contained chapters, Halacy takes up the side effects of the energy trap (pollution and blackouts), recounts the development of our fossil fuels, examines fission and fusion as doubtful alternatives, and considers the possibilities of solar energy. In places he goes beneath the surface to make some seldom-raised points about the real cost of generating plants fired by different energies, and related issues. Clearly written and fact-filled, this work will be useful for research papers.

16. Pringle, Laurence. **Energy: Power for People.** New York: Macmillan, 1975. 147pp. $6.95. Gr. 7-12.

A nicely illustrated overview of the American energy situation, this work explains the "energy crunch" in separate chapters on the use or potential use of coal, oil and natural gas, and alternative energies. A chapter on energy from the atom includes a statement that "there is not yet any plentiful source of truly clean energy—only a choice of poisons." Suggestions are made for conservation and increased efficiency in industry, housing, transportation, and consumer habits; young people are urged to take citizen action.

17. United States Executive Office of the President, Energy Policy and Planning. **The National Energy Plan.** Washington, D.C.: U.S. Govt. Print. Off., 1977. Dist.: U.S. Dept. of Energy, Technical Information Center. free. Gr. 7-12.

This publication sets forth the Carter Administration plan for the present, the near future, and the long term in every energy area. Nine chapters take up the origins of the U.S. energy problem; the continuing crisis; principles and strategy of the plan with regard to conservation and energy efficiency, oil and natural gas, coal, nuclear, and hydroelectric power, and nonconventional sources; the role of government and the public in the plan; and the outlook for the future. Three main objectives of the plan are reiterated: to reduce dependence on foreign oil now; to keep oil imports sufficiently low in the near future; and to have renewable and "essentially inexhaustible" sources of energy in the long term. There is no index but a good table of contents. An indexed version is available from Ballinger Publishing Co. for $10.95 or $6.95 paperback. This title is essential for every reference collection and for many social studies teachers.

18. Butler, S. T., and Raymond, Robert. **The Energy Explosion.** Frontiers of Science. Garden City, N.Y.: Anchor Pr., 1976. 84pp. $2.95 pap. Gr. 8–12.

More than 40 rather complicated energy concepts are each presented here in black-and-white comic-strip format for the reluctant reader or even the more interested reader. Topics include the revival of the Stirling engine, the problems of breeder reactors, fusion by laser, and flying on hydrogen. This is the print equivalent of a collection of single-concept film loops and is well done.

19. Goldstein, Eleanor C., ed. **Energy.** 2 vols. Gaithersburg, Md.: Social Issues Resources Series, 1977. unp. Vol 1, $50. Vol. 2, $20. Gr. 9–12.

These two volumes are part of SIRS' series on topical social issues. Volume one includes 100 articles from newspapers, magazines, journals, and reports on various aspects of the energy crisis including proposed solutions, alternative energy sources and their effects, energy conservation, and the social and political consequences of the crisis. It begins with a report of the 1971 hearings before the Senate Subcommittee on Minerals, Materials and Fuels; remaining articles were published from 1972 to 1974. Volume two includes 40 articles published in 1975 and 1976; eleven of these articles focus on nuclear energy. The articles are presented chronologically, rather than by subject. With the exception of some of the newspaper articles, the articles require a secondary school reading level. The loose-leaf format allows many students to use each volume simultaneously; articles are color-coded to make reassembly convenient. The particular value of these volumes lies in the fact that they include articles from journals and reports such as *Panhandle*, a publication of Panhandle Eastern Pipeline Company, which would not normally be received by school libraries. Recommended as a ready-reference tool for reports, debates, and class discussion, these works do not provide in-depth coverage of any particular topic; rather, they offer a broad, relatively recent overview of the energy situation.

20. Hayes, Denis. **Energy for Development: Third World Options.** Worldwatch Paper No. 15. Washington, D. C.: Worldwatch Institute, 1977. 44pp. $2 pap. Gr. 9–12.

In this clearly written paper, Hayes contends that alternate energy sources such as sunlight, wind, water, and biological sources make more sense economically for Third World countries, which are still heavily rural, than they do for industrialized nations. He sees the Third World's lack of coal as an ironic advantage which will enable them to move directly into the solar era, and he believes that development of nuclear power would be a costly mistake that these financially strapped nations should avoid. Citing examples from international research reports, he advocates decentralized technologies which are economical and related to local needs. This book could be useful in social studies classes studying the development of non-Western nations.

21. Kaplan, William, and Lebowitz, Melvyn. **Energy and Fuels.** Student Scientist Series. New York: Rosen, 1976. 137pp. $7.97 plb. Gr. 9–12.

This is one of a series written for secondary students who want to explore scientific topics in greater depth. Early chapters discuss the need for and the nature of energy. Remaining chapters focus on coal, oil and gas, nuclear fuels, and energy and the environment. The closing chapter on the energy crisis tends to oversimplify complex issues. Sample "questions to think about" follow each chapter. This book is probably most useful for general science students whose knowledge of energy is limited. The chapter on nuclear fuels, by nature a more complex topic, is particularly well presented.

22. Sierra Club. **Energy Packet.** San Francisco, Calif.: Sierra Club Information Services, n.d. $1. Gr. 9–12.

This packet of materials from the national environmental organization includes "Energy and the Sierra Club," which states the club's position on national energy decisions, and "Nuclear Power and the Sierra Club," which outlines the club's position on the development of nuclear power. The packet also includes reprints from the *Sierra Club Bulletin* such as "Solar Energy Now," "Geothermal Energy, Prospects and Limitations," and "700,000,000,000 Barrels of Soot."

23. Udall, Stewart; Conconi, Charles; and Osterhout, David. **The Energy Balloon.** New York: McGraw-Hill, 1974. 288pp. $7.95. Penguin, $2.95 pap. Gr. 9–12.

Former Secretary of the Interior Udall and his co-authors recall that in the post-World War II era Americans believed that massive oil strikes abroad represented their own oil. The authors warn that offshore oil reserves and alternate sources of energy will not begin to meet future energy needs, and they feel that "the primary task now is to develop a frugal national lifestyle that conserves and husbands oil supplies to prolong the petroleum age as long as possible." They are not sanguine about any alternate sources of energy except small-scale solar and wind installations. Specific chapters call for an industrial reformation, restoration of transportation options, and a reshaping of urban and suburban environments to conserve energy on a massive scale. The book provides a social and political history of energy in the United States. Taking a popular rather than a scientific approach, this work will be most useful for social studies assignments.

24. Wilson, Mitchell. **Energy**. rev. ed. Life Science Library. Morristown, N. J.: Time-Life Books, 1970. 200pp. $5.94 Kivar; $7.65 plb. Gr. 9–12.

This volume of the Life Science Library is a comprehensive, heavily illustrated book which examines energy from Galileo's discoveries to the laser beam. Sources of energy, measurement of energy, and each form of energy—mechanical, electrical, radiant, chemical, and heat—are discussed with the clarity typical of this series. One picture essay surveys the history and production of oil; other essays cover electrical phenomena, the power of the atom, and future power sources. A chronology recounts the history of atomic energy. For advanced junior high readers, and senior high students.

25. Clark, Wilson. **Energy for Survival: The Alternative to Extinction**. Garden City, N.Y.: Anchor Pr., 1974. 652pp. $12.50. Gr. 10–12.

Clark surveys the energy scene, as of 1974, in a comprehensive, heavily researched report covering nearly every aspect—from corporate energy monopolies to bioconversion and decentralized power systems. A fourth of the book is given over to solar. Long but nontechnical and decently written, this title presents the case for the development of "more benign alternatives" than nuclear fission. It could be used for reference purposes in both science and social studies since few secondary students are liable to tackle it all.

26. Considine, Douglas M., ed. **Energy Technology Handbook**. New York: McGraw-Hill, 1977. 1884pp. $49.50. Gr. 10–12.

Although most of the material in this handbook is technically oriented, it can be a useful source for the student seeking information about energy sources. It lists estimated reserves, production levels, existing and planned generating facilities, and information about exploration. The book is divided into nine sections: Coal Technology, Gas Technology, Petroleum Technology, Chemical Fuels Technology, Nuclear Energy Technology, Solar Energy Technology, Geothermal Technology, Hydropower Technology, and Power Technology Trends. Each section has its own detailed table of contents, should the general index not suffice. Typical major headings under "Coal," for example, are: Properties, Reserves and Testing; Safety and Environment, Coal Preparation and Handling; and Coal Conversion Processes. Major headings under "Nuclear" include: Nuclear Fission Reactors, Nuclear Power Plant—Concept to Startup, Thermal Reactors, Fast Breeder Reactors, Fusion Power, and Nuclear Fuels.

27. Ford Foundation Energy Policy Project. **Exploring Energy Choices**. New York: Ford Foundation, 1974. 141pp. $.75 pap.; $.60 ea./50 or more copies. Gr. 10–12.

The first published product of the Ford Foundation's Energy Policy Project, this is a shorter, preliminary version of *A Time to Choose* (see No. 173).

28. Heppenheimer, T. A. **Colonies in Space**. Harrisburg, Penn.: Stackpole Books, 1977. 224pp. $12.95. Gr. 10–12.

Three of the 16 chapters in this book concern the energy promise in space. The author, a planetary scientist with a doctorate in aerospace engineering, dismisses fusion and the breeder reactors as presenting insoluble technical problems, and he points out that ground-based solar power is too expensive. He makes a case for power satellites—maintaining that these satellites can be built in a space colony at a lower cost than building conventional generating plants on earth. Chapters three and four give the technical details. Well-documented, the book also examines the social implications of space colonization.

29. Metzger, Norman. **Energy: The Continuing Crisis.** New York: Crowell, 1977. 242pp. $12.95. Gr. 10–12.

A literate, somewhat pessimistic view of energy supply and the control of its demand which also examines the crisis and what can be done about it. Offering no technological panaceas, the author, a science writer, recounts the background and potential of each power source (although coverage is uneven) and the factors affecting the success and failure of each. He sees the main issue as whether the government and private industry can properly prepare for new technologies and renewable energy sources. Nontechnical but very detailed, this book is for the serious senior high social studies student.

30. Mott-Smith, Morton. **The Concept of Energy Simply Explained.** New York: Dover, 1964. 215pp. $2.75. Gr. 10–12.

Formerly titled *The Story of Energy*, this work explains how energy is used today, as well as the history of energy. It is geared for secondary physics or chemistry students who are working on outside assignments. Chapters on the conservation of energy and entropy are particularly good as supplementary reading.

31. National Science Teachers Association. **Factsheets on Alternative Energy Technologies.** Washington, D.C.: National Science Teachers Assn., 1977. Dist.: U.S. Dept. of Energy, Technical Information Center. 4–8pp. ea. free. Gr. 10–12.

Produced by the National Science Teachers Association under contract with ERDA, each of these 19 factsheets provides a comprehensive introduction to an energy source, the resource base, its technology, environmental problems accruing from it, and the economics of and prospects for that source. Six tackle the topics of energy conservation, conventional and breeder reactors, energy storage, technology, and environmental impacts of alternative energies. The last two provide a glossary of energy terms and a bibliography of energy sources. An effort is made to place each source in the total energy picture, and references are included. The factsheets are recommended for ease of use, clarity, and organization. Their vocabulary and technical content will restrict them to senior high level. The titles are: 1. "Fuels from Plants (Bioconversion)," 2. "Fuels from Wastes (Bioconversion)," 3. "Wind Power," 4. "Electricity from the Sun I (Solar Photovoltaic Energy)," 5. "Electricity from the Sun II (Solar Thermal Energy Conversion)," 6. "Solar Sea Power (Ocean Thermal Energy Conversion)," 7. "Solar Heating and Cooling," 8. Geothermal Energy," 9. "Energy Conservation

(Homes and Buildings)," 10. "Energy Conservation (Industry)," 11. "Energy Conservation (Transportation)," 12. "Conventional Reactors," 13. "Breeder Reactors," 14. "Nuclear Fusion," 15. "New Fuels from Coal," 16. "Energy Storage Technology," 17. "Alternative Energy Sources (Environmental Impacts)," 18. "Alternative Energy Sources: a Glossary of Terms," 19. "Alternative Energy Sources: a Bibliography."

32. Primack, Joel. **Energy**. Vital Issues Series. Washington, Conn.: Center for Information on America, 1977. 4pp. $.45. Gr. 10–12.

The author, a physics professor at the University of California at Santa Cruz, discusses the economic factors affecting the coal vs. nuclear energy debate, and the environmental and social impact of both energy sources. He suggests a compromise on using nuclear energy: avoid the long-term commitment implicit in breeder technology, but build a limited number of nuclear plants to supplement the energy supply through the next several decades of shortages until alternative sources can be developed.

33. Sheridan, David, and Gordon, Irene S. **Alternative Energy Sources**. Vital Issues Series. Washington, Conn.: Center for Information on America, 1975. 4pp. $.45. Gr. 10–12.

This paper describes briefly the state of research and funding for solar, geothermal and fusion energy, tidal power, fuel cells, and total energy systems. The authors conclude that funding and legislation to develop "clean new energy technologies" falls short of a major effort. Could be used in both science and social studies classes.

34. Speckhard, Roy A. **Basic Facts on the Energy Crisis**. Homewood, Ill.: Learning Systems Co., 1974. 81pp. $2.85 pap. Gr. 10–12.

Intended for individuals who want to expand their knowledge in fields outside their own field, this is one of a series of "Plaids," programmed learning aids, prepared by textbook authors. Information is divided into five broad areas: man and energy, an overview; energy and the evolution of human society; energy resources, social conditions, and the economy; energy and politics; and energy and the future. Tests are included throughout the work; there is also a final exam covering the five areas. The tone is somewhat like a textbook. This will probably be most useful with social studies students working on independent study projects.

35. Clegg, Peter. **New Low-Cost Sources of Energy for the Home**. rev. ed. Charlotte, Vt.: Garden Way, 1977. 252 pp. $8.95; $6.95 pap. Gr. 11–12.

The five sections of this book—solar energy, wind power, water power, water/waste systems, and wood heating—are organized along parallel lines. Each covers the principles of the form of energy; its availability; cost; the method of incorporating it into the design of the house; additional references; and a catalog reproducing advertisements from manufacturers in each field (not always with addresses.) The ads dominate the book, but the sections which discuss principles are useful; other portions tend to be cursory.

36. Edison Electric Institute. **The Transitional Storm.** New York: Edison Electric Institute, 1977. 78pp. $.75; discount available on bulk orders. Gr. 11–12.

This is one of three titles in the Decisionmakers Bookshelf produced by the major association of America's investor-owned electric utility companies. The initial article defines the storm as an "agonizing period between two great fuel epochs: the present fossil fuel epoch, which, except for coal, has already begun phasing out, and a new epoch, which, except for uranium, hasn't yet been precisely identified." The remainder deals with the impact of this storm on the economy, the question of a no-growth economy vs. economic expansion in terms of current energy problems, and national energy policy. A statement by the Board of EEI spells out very specific recommendations for national energy policy. This work provides an interesting presentation of the utility companies' point of view. Recommended for sophisticated high school juniors and seniors.

37. Hand, A. J. **Home Energy How-To.** Popular Science. New York: Harper, 1977. 258pp. $9.95. Gr. 11–12.

One third of this *Popular Science* book is devoted to conserving energy, the remainder to producing it. Sixty pages are devoted to solar energy, somewhat fewer to wind and wood, and fewer still to water and biofuels. Enough information is provided, however, to build an experimental windplant, and sources of additional information (with addresses) are noted for each energy source. Chapters two through seven cover insulation, caulking and weatherstripping, windows and doors, conventional heating systems, home cooling, and architectural considerations for conservation. Well-indexed, this is a good reference tool for carpentry, electricity, or heating and air conditioning classes.

38. "Power Plant Performance: Nuclear and Coal Capacity, Factors, and Economics." **Council on Economic Priorities Newsletter,** Nov. 30, 1976. New York: Council on Economic Priorities, 1976. 6pp. $1; inquire for multiple copy prices. Gr. 11–12.

Based on a longer study, *Power Plant Performance* by Charles Komanoff, this comprehensive newsletter issue maintains that an accurate assessment of nuclear and coal efficiency and cost is essential to U.S. energy planning. Among its findings are that coal is competitive with nuclear power for new plants in the Northeast and up to 20% cheaper than nuclear power elsewhere. Nuclear operating reliability will average 45-60% (compared to 60-70% for coal), rather than the 70-80% projected by the nuclear industry and the government. CEP suggests that the nuclear industry revise its inflated factor projections to reflect reality, so that the public can make an informed decision on the expansion of nuclear capacity and the real savings in oil it will afford. This newsletter could be used in social studies classes studying economic problems with political implications.

39. Stoner, Carol Hupping, ed. **Producing Your Own Power: How to Make Nature's Energy Sources Work for You.** Emmaus, Penn.: Rodale Pr., 1974. 322pp. $8.95; Vintage, $3.95 pap. Gr. 11–12.

This guide describes efforts to generate small-scale electric power or to heat homes or water by energy sources other than fossil fuel. Specialists discuss the pros and cons of wind power, water power, methane, wood, solar energy, and combined energy systems. The section on water power includes instructions on how to build small dams and an hydraulic ram. The appendix gives practical suggestions for conserving energy in existing structures. The focus of the book is very technical, but the clear writing style makes it easy to understand. It will be of most use to teachers and students in conservation programs, and to those who are seeking off-beat power sources.

NONPRINT

40. Energy and You. 6 filmstrips w/6 cassettes or 3 discs. 56–68 fr., 7–12 min., color, w/tchr's. guide. Society for Visual Education, 1975. $83 set; $11 ea. strip; $7.50 ea. cassette or disc. Gr. 1–6.

This series of six sound filmstrips introduces the viewer to energy: what it is, where it comes from, how it works, where it goes, and how it is stored. The first two strips are suitable for primary grade students, the remaining four are better suited to third grade and up, although the same characters converse in each strip and each strip builds on the preceding ones. They should be shown in sequence for optimal learning. The series portrays Lucinda and a blue-faced superman called Big E (for energy). Kids who are used to Saturday morning television's unsubtle art and humor will probably feel comfortable with the characters and the set. In Part I, *Energy: What Is It?*, Lucinda meets Big E, who explains what energy does in rhyme like "The sun, the sun, That's where it all comes from," or "You can't see energy— it's like the wind. It makes things move, from beginning to end." Relatively low-key, the strip introduces the concept that energy is present whenever things move, grow, or change, and cites examples of energy familiar to younger students. In Part II, *Energy: Where Does It Come From?*, Lucinda gets the word on two points: how the food energy in her hamburger comes originally from sun energy, and where a car gets its energy to move. Big E gives a capsulized version of how oil and natural gas were formed, how oil is refined, and how energy is released when gas is burned in a car. Coal and its origins also get a brief mention but the real plug is for the sun as the ultimate energy source. Part III explores the question *Energy: How Does It Work?* Big E and Lucinda go bowling and learn that everything that moves is doing work and that in a bowling alley the pinsetting machines do work. Big E discusses gravity and potential vs. kinetic energy—with coal and potato chips used as examples of potential energy turned into kinetic. Lucinda reviews the findings in the first three strips in Part IV, *Energy: Where Does It Go?*, and wonders indeed where the energy goes. Big E takes her to a power station and explains how chemical energy is turned into heat energy, then mechanical energy, and finally electrical energy, which in homes becomes light and sound energy. There is a recap halfway through and a few frames at the end on atomic energy, but the overall message is that energy is never lost, simply changed. Part V, *Energy: How Do You Store It?*, examines the concept of energy storage through the example of a car's battery. Human energy's derivation from food's stored energy, and the storage of energy in radioactive materials such as uranium and radium, are introduced. Ten frames describe how solar energy can be utilized and the problems of storage and diffuseness associated with it. Lucinda is enthusiastic about its potential. In Part VI, *Energy: Using It Wisely*, Big E visits Lucinda's house and finds examples of energy waste at every turn: lights left on in empty rooms, a brother taking long hot showers with the radio on, the refrigerator door open, two cars and a motorcycle, etc. His practical advice on how to save energy and money before energy supplies run out winds up the series.

41. First Films on Science: Energy and Motion. videocassette, 13 min., color, w/tchr's. guide. Mississippi Authority for Educational Television, 1975. Dist.: Agency for Instructional Television. $110; $25/7 days rental. Gr. 2–4.

This short tape introduces some basic energy concepts: energy is used whenever anything moves or is moved; heat is a form of energy; energy goes from one place to another; sound is a form of energy; electrical energy is changed to other forms of energy; and chemicals can give us energy. The terms primary energy (energy from the sun) and secondary energy (energy that is formed as a result of primary energy) are defined. The tape is well paced for the younger student and uses simple examples of energy changes (hitting a nail, a tug o'war, a balloon let go, toy trains on a track, wood burning) to illustrate the concepts. Three other tapes in this series—*Solar Energy, Changes in Classes of Energy,* and *Changes in Kinds of Energy*— should be considered but were not available for preview at this time.

42. Forms of Energy. 1 filmstrip w/1 cassette or disc, 44 fr., 10 min., color, w/tchr's. guide. Society for Visual Education, 1977. $18.50; $11 filmstrip; $7.50 cassette or disc. Gr. 2–5.

Two impressionable kids, Basil and Bernice, go to the movies and cross over into a haunted house where they meet Energenie who, like energy itself, is invisible. The kids think Energenie is crazy, but she does teach them about heat energy, light, what electricity does, and how important the sun is, all while whipping up a new stew for them. Seven review frames with questions printed on each frame provide a useful recap of this introduction to the idea of energy forms.

43. Matter and Energy (Series). 6 filmstrips w/6 cassettes or 3 discs. 48–57 fr., 10–12 min., color, w/tchr's. guide. Clearvue Inc., 1976. $72.50 ser.; $15.95 ea. Gr. 2–5.

A series which explores the basic concepts of matter and energy through an interchange between a comic-opera German professor, Whizbang, and his robot, Rodney. The complexities of molecular theory, the speed of light, the principles of simple machines, and the properties of magnetism, electricity, and sound are covered. The cartoon artwork is above average and the length is good not only for its intended primary-intermediate audience, but also for middle school classes and slower junior high classes.

44. Energy: A First Film. 16mm, 8 min., color, w/tchr's. guide. BFA Educational Media, 1970. $130; $15/3 days rental. Gr. 4–6.

A brief, basic energy film which makes four main points: energy exists in many different forms, such as motion, light, and chemical energy; energy may be changed from one form to another, e.g., the motion energy of a spinning turbine can be changed to electrical energy; energy can be stored, as it is in logs which are later burned to provide heat and light; most energy begins as light energy from the sun. In making the last point, the function of the sun in the food cycle is left somewhat hazy. These are broad concepts, for the most part deliberately and clearly stated. The use of such words as "turbines" and "photosynthesis" precludes its use with primary grade levels.

45. Energy and Its Forms. 16mm, 11 min., color, w/tchr's. guide. Coronet Films, 1961. $160. Gr. 4–6.

The film is concerned with eight different forms of energy—movement, position, light, sound, heat, electric, chemical, and nuclear—and with the fact that one form of energy changes to another form whenever we use it. It traces almost all forms of energy, including oil and coal, to the light energy of the sun. Commonplace examples of energy forms are used.

46. Energy: Where It Comes From, Where It Goes. 1 chart, color. Gulf Oil Corp. free as long as supply lasts. Gr. 4–6.

Against a black background, a brightly colored American eagle points to percentages showing the three major sources of energy in the U.S. and the three main ways it is used. The chart accompanies the pamphlet "An Eagle Eye on Energy" (see No. 4).

47. Introduction to Energy and Engines. 1 silent filmstrip, 31 fr., color. Standard Projector & Equipment Co., 1966. $7. Gr. 4–8.

The following topics are developed in this strip: energy and what it is; the various sources of energy, historical and current; how the use of various forms of energy has changed over the years; how man uses energy to make a better world. The filmstrip provides an overview of how man used his own energy and later the energy in gunpowder, coal, wood, and oil to do his jobs. It takes the viewer through the invention of the steam engine and the internal combustion and jet engines to nuclear energy and man-made lasers. A vocabulary frame precedes the content; the last four frames include two tests and their answers. The illustrations are sketchy.

48. This World of Energy (Series). 5 filmstrips w/5 cassettes or 5 discs, 47–57 fr., color, w/tchr's. guide. National Geographic Society, 1974. $74.50 ser. Gr. 4–8.

Energy in the Earth, the first part of this series, explains that the sun is the source of all energy, including food energy, and describes derivative sources which provide the energy to do our work. *Using Energy* traces the history of man's energy use and particularly our dependence upon fuels to produce food and do our work, and notes America's vast appetite for electricity compared to other nations. *Fossil Fuels* explains how these fuels are formed, extracted, processed, and used to provide the energy we demand and tells what is being done to find new fossil fuel sources and to extend the current supply. *Nuclear Power* gives a basic introduction to the principles of nuclear energy and points out the danger of thermal pollution and radiation involved in generating nuclear power. *Energy for the Future* discusses energy shortages, energy waste, and potential energy sources. Fusion, geothermal, and solar power are shown, but the emphasis is on the potential of solar energy. This recapitulates some points made in the earlier strips. The series is well explained as far as it goes, but it is superficial because of the coverage and length. A glossary of nuclear energy terms might be provided to the student before showing the fourth strip. The five parts can be used with large or small groups for independent study; each part is color-coded.

49. Energy and Living Things. 16mm, 10½ min., color. Centron Educational Films, 1972. $165. Gr. 4–9.

In a somewhat circuitous fashion, this film describes the chain of plants converting light energy into chemical energy in the form of food; animals eating this food but not utilizing all of it, so that uneaten or undigested plant material is eaten by scavengers or broken down by decomposing organisms, or, if left long enough, converted by high pressure and temperatures into fossil fuels such as coal and oil. The energy in the plant and animal material remains as stored energy. The film concludes with a tabletop demonstration showing that if fuel is burned faster than it is added, the burning light must inevitably go out. It asks the question, "What will happen when our supply of coal and oil run out?"

50. Energy. 150 cards. Scholar's Choice Ltd., 1977. $8.95. Gr. 5–7.

This "activity box" contains 150 cards which provide teachers with strategies for demonstrating to students what energy is and how we rely upon it. The first of five sections, "Energy—What Is It?," is intended for science classes. It introduces, on separate cards, the ideas of work, force, measuring work, power, and forms of energy. The remaining four sections are interdisciplinary with a strong social studies emphasis. They are devoted to present energy sources, how we use energy, methods of conservation (at home, in transport, in school, and in the community), and alternative sources. Suggestions for investigation, discussion, or activities are included on many cards. The complexity of the cards and activities varies; some cards could be used independently, while others will need additional discussion or supplementation. Essentially a resource tool, this could be used by teachers who are asked to put together an energy packet or unit in environmental studies programs, or shared by science and social studies teachers.

51. Energy: Can't Do without It. 16mm, 14 min., color, w/tchr's. guide. McGraw-Hill Films, 1974. $225; $12.50 rental. Gr. 5–8.

The film achieves only two of its stated behavioral objectives: viewers will recognize the extent to which energy is needed in our everyday lives and appreciate the relationship between energy and jobs, and energy and leisure. They will not, however, have more than a superficial awareness of the limitations of depletable resources nor will they "realize the need to use present sources of energy more efficiently now and in the future." The pluses for the film are its explanation and familiar examples of energy, work, and fuel. It also shows that cars and buildings are energy guzzlers, e.g., the World Trade Center uses as much energy in an eight-hour day as the city of Schenectady in 24 hours. In the guise of being open-ended, it leaves today's plugged-in kids with the awesome question, "Imagine what life would be like without it, " but provides no serious suggestions for altering our wasteful lifestyle or finding other sources of energy. The contrast the film offers with a Latin American family using a plow drawn by animals is too extreme to have much carry-over. The post-film projects, including collages of essential, nonessential, and luxury energy uses, are generally weak.

52. Energy Sources of the Future. 16mm, 15 min., color, w/tchr's. guide. McGraw-Hill Films, 1974. $240; $12.50 rental. Gr. 5–8.

This film gives a very quick overview of ten alternate sources of energy and conveys the idea that people must decide which sources shall be developed to meet energy needs. Those included are shale oil (one of the few films which shows the shale actually being processed), underground coal gasification, repulsive magnetic levitation to run trains, superconductivity, nuclear power, solar energy, wind, steam, tides, and garbage. The plant in Brittany which uses tidal power to generate electricity is shown, as is the French solar furnace, but the film is honest about the current usefulness of tides, wind, and geothermal and solar energy. The moral: "The decisions we make today will determine whether the lights go on tomorrow." It is geared more to social studies than science classes since the idea of citizen responsibility is stressed.

53. California's Energy Resources. 1 filmstrip w/cassette, 75 fr., color. Academy Films Distribution Co., 1977. $25. Gr. 5–9.

Much of this strip is devoted to showing specific examples of power plants in California and explaining how hydroelectric and steam generating systems work. It includes a good description of how a generator works. The operation of nuclear-powered generating systems is briefly described, as are the steam wells in Sonoma County which yield geothermal power. It concludes with examples of how solar energy may be used. Although it does not assess California's nonhydroelectric resources, this may still be useful to California students. The visuals are not up to the narrative, which is well paced.

54. Energy. 8 posters, 12½" x 17½", color. Educational Insights, 1977. $7.95. Gr. 5–9.

Eight four-color posters bound in a folder deal with oil and gas, hydroelectric power, coal, nuclear power, solar energy, wind energy, and geothermal energy, as well as future energy prospects. On semi-stiff board, each poster demonstrates how the source of energy is obtained and utilized. Since there is a wealth of information about each type of energy—its history, advantages and disadvantages, current prospects, and an activity demonstrating its principle—printed on the front and back of each poster, the set may be more useful as a pass-around item or a ready-reference source than it would be tacked on a bulletin board. The inside cover of the folder contains a teacher's guide with suggestions for discussion and follow-up activities and for integrating the study of energy into the curriculum. Also included is a list of outside sources which can be contacted for further information on the general topic of energy.

55. The Wide World of Energy (Kit). 1 filmstrip w/cassette, 73 fr., 8 min., color; 1 energy game; 10 experiment cards; 18 color study prints, 13" x 18"; 30 comic books; tchr's. guide. Walt Disney Educational Media Co., 1972. $119. Gr. 5–12.

This kit has a variety of audiences. It contains 30 copies of a comic book, *Mickey Mouse and Goofy Explore Energy,* which gives an inadequate explanation of the difference between fission and fusion. A standard board game on energy for up to six people contains cards which can also be used as flash cards. Ten energy experiment cards are included. A somewhat superficial sound filmstrip, *The Search for Power and Industry,* tells in comic-strip form how man discovered and harnessed energy and implies that atomic energy is *the* answer to the energy problem. Eighteen energy study prints, arranged in three categories, have bright but banal illustrations on one side, and substantial information and diagrams on the reverse. The prints in *The Nature of Energy* packet could be used by grades 6–7; those in *Man Puts Energy to Work* by grades 7–8. The third group, *Energy Doorways to the Future,* contains information on such topics as radar, fuel cells, lasers, and masers which is more suited to high school physics students than bright middle-school students. The unbalanced ability-range of this material, plus the actual content—only one fairly average sound filmstrip along with comic books and an inexpensive board game—make this a marginal purchase.

56. Energy Carol. 16mm, 10:33 min., color. National Film Board of Canada, 1975. $160; $20/1 day rental. Gr. 6–12.

A fabulous tale based on Dickens's Christmas lesson. In this cartoon version Ebenezer Scrooge is president of Zeus Energy Company, whose motto is: "If we didn't waste, we couldn't grow." Scrooge is visited by three spirits showing him energy in the past, as it is today (with the air conditioner and furnace going simultaneously), and as it will be in the future when the fossil fuels are depleted. Scrooge had planned on being deep frozen and, later, defrosted to live again. One look at his own splashily lit mausoleum after the power goes off convinces Scrooge that he must never waste again. In great National Film Board style, this is a humorous film lesson about things not so humorous, like the way we live and what the future may hold. Fast, sometimes hard-to-catch dialogue limits it to the sixth grade and up as a discussion starter in both social studies and science classes.

57. We Will Freeze in the Dark. 16mm, 42 min., color, w/tchr's. guide. McGraw-Hill Films, 1977. $630; $65/1 day rental. Gr. 6–12.

An up-to-date, attention-grabbing documentary which features Nancy Dickerson as the guide through a land where energy conservation methods are strictly enforced. The film opens with two sinister men denuding a typical home of small appliances and padlocking the larger appliances, a portent of what might happen if Americans continue their energy-wasteful habits. Dickerson covers a lot of territory, from the bright lights of night horse-racing at Meadowlands to the lowered lights in Hardee's hamburger chain, and talks with a major auto designer about energy conservation. The film contends that everyone has talked about energy conservation since 1952, but no one has done anything about it, e.g., 1985 guidelines may be set for fuel consumption by cars, but there are still no federal efficiency requirements for refrigerators. The film includes footage of Presidents Nixon and Ford, representatives of the Illuminating Engineering Society, and oil and chemical company executives. There are some positive notes: a refrigerator manufacturer stating that savings can be effected; architects and designers showing off an elementary school in Reston, Va., which utilizes student body-heat as well as solar collectors to heat the building, and which was financed in part by a grant from the King Faisal Foundation in Saudi Arabia. All agree that the era of cheap fuel is over and we must conserve energy while seeking long-term methods of achieving energy independence. As slickly produced as the 7 o'clock news, this is best adapted to social studies classes from middle school up.

58. Energy for Tomorrow (Series). 3 filmstrips w/3 cassettes, 56–70 fr., 12–14 min., color, w/tchr's. guide. Educational Materials and Equipment Co., 1976. $59.95 ser. Gr. 7–9.

Part I, *Energy Alternatives,* stresses the advantages of water as an energy source and looks at river flow, wave motion and tides, and temperature differences in oceans. It also considers the pros and cons of wind power, geothermal energy, and energy from bio-conversion (the conversion of organic wastes such as cow manure or sewage into methane), and man's need to live in harmony with the earth. Part II, *Solar Energy,* develops an understanding of where heat comes from, how the sun's light is absorbed and used by plants, and the part played by the sun in the formation of fossil fuels and in water power. The problems of gathering and storing the sun's energy to make practical use of it are covered. Part III, *Nuclear Energy,* explains both the fission and fusion processes and presents a good overview of how nuclear power is produced, what its potential is, and the problems impeding its widespread use. This series provides a good summary of energy possibilities and provides a basis for more detailed research.

59. Energy: Here Today—Gone Tomorrow? (Series). 4 filmstrips w/2 cassettes, 47–54 fr., 6:31–8:07 min., color. Eye Gate Media, 1974. $49.80 ser.; $10 ea. filmstrip; $7 ea. cassette. Gr. 7–9.

In Part I, the American home is likened to a spaceship in that it is a protected environment which requires energy to be built and maintained. The second strip shows the low-energy self-sufficient lifestyle of the colonists and notes that today Americans cannot be self-sufficient. It shows that our homes consume vast amounts of energy as does our system of transporting, processing, and distributing food. Part III describes the evolution of the energy sources we now command, questions our dependence on oil imports, and asks what will be the future social, economic, and health effects of changes in energy-consumption patterns, e.g., a switch to smaller cars. The final strip lays the blame for the energy crisis on the public's increasing appetite for goods and conveniences, as industry uses more and more energy to produce and deliver them. It suggests a tri-part solution: find immediate methods of reducing energy requirements, find new sources of fossil fuels, and develop nonfossil fuel sources of energy. There are very specific suggestions for industrial and home conservation, but the sections on finding new sources of traditional fuels and developing alternative fuels are both slighted. Frames at the end of each part review the content and suggest post-viewing activities. There is no formal teacher's guide. Although this is an overview set which could be used with upper elementary students, its very rapid narration and coverage of a wide range of topics make it more likely for junior high students.

60. The Energy Crisis. 160 slides w/2 cassettes and 2 discs, color, w/tchr's. guide. Educational Design, 1973. $99. Gr. 7–10.

Opening with a slide of New York City during the 1965 blackout, Part I of this program traces the history of the energy problem from man's discovery of fossil fuels to the present fuel shortage. It examines the nature and uses of fossil fuels and possible alternate power sources for the future, opening up the question of how we will plan for the future. Part II presents the solutions offered by disparate people: a scientist who sees the need for a no-nonsense, heavy-research approach; a businessman who believes industrial energy needs and the economy should come first; a confused recent college graduate who likes her appliances; an environmentalist who doesn't want a nuclear generating plant in her town because of the dangers from pollution and radiation; a Congressman who prefers self-regulation to government controls; a diplomat who points out that the U.S. and other industrial nations have found that energy resources are political resources. The set covers the options but does not go deeply into any of them. The emphasis is on the need for "action now."

61. Power without End. rev. ed. 16mm, 16 min., color. Xerox Films, 1977. $270. Gr. 7–10.

Asserting that we are using up fossil fuels a million times faster than nature can create them, this film offers a quick, attractive introduction to the alternatives. Among the energy producers and users covered are: the solar furnace in France, designed to melt metals and produce hydrogen and oxygen; solar cells; solar housing in New Mexico; solar heating and cooling on a New York phone company building; the possibility of solar space stations; desert collectors. Wind power, geothermal energy (with its ecological problems), hydroelectric power (and the claim that the Aswan Dam has played havoc with agriculture in the area below it), organic waste, and fusion power are all covered briefly. Viewers may differ with the film's conclusion that these sources "are capable of supplying us with our total energy needs forever."

62. Alternative Energy Sources. videocassette or 16mm, 30 min., color. Indiana Univ. Audio-Visual Center, 1972. $220 videocassette; $315 16mm; $12.50/3-5 days rental. Gr. 7-12.

Opening with the dynamiting of a coal field in northern Arizona to meet the growing energy needs of the Southwest, this film first looks at the destruction of hundreds of square miles in the West caused by stripmining. Narrated by Robert Redford, it shows coal slurry pipelines running hundreds of miles, involving huge amounts of water and the overtaxed Colorado River. Posing the question of whether we can afford to dig up and pollute the West to satisfy our energy habits, it describes damage already being done to plants and trees and to 27 national parks, recreation areas, and monuments. The pros and cons of alternative energy sources are considered. U.S. Department of Energy Secretary Schlesinger speaks for the breeder reactor; Dr. John Gofman endorses fusion. This film concludes that serious scientists see the logic and necessity for alternative energy sources rather than continuing pursuit of bigger fossil-fueled power plants, and it exhorts citizens to support the alternatives.

63. Energy (Series). 8 filmstrips w/4 cassettes or 4 discs, 50-65 fr., 8:23-11:05 min., color, w/tchr's. guide. Clearvue Inc., 1974. $99.50 ser.; $15.95 ea. Gr. 7-12.

Part I presents a rather superficial overview. Part II discusses radiant energy, explaining the use of solar cells in converting sunlight to electricity and emphasizing the problems of solar power, such as the huge collecting areas and storage problems. Strips III and IV focus on chemical energy sources. The former describes how oil is produced, secondary recovery, and products derived from oil; the latter notes that coal is the giant of fossil fuels and predicts that strip mining will be a prime energy source for the next decade, but will soon exhaust the coal supply. Part V gives the history of our knowledge of electrical energy and how electric power is obtained. Strip VI contains a good section on the conversion of mechanical energy to electric energy in a hydroelectric plant. Part VII explores atomic structure and fission, giving a very brief description of a reactor, but it does not deal with attendant hazards. The last strip shows what the economic and social effects of an energy shortage could be, and urges conservation as well as a search for new sources. It also touches on lasers, atomic fission and fusion, geothermal energy, lunar energy, wind power, burning coal underground, liquefaction of coal, oil shale, and oil from the ocean floor. Its conclusion presents a social problem: just as important as finding energy is the way energy is used, i.e., for peaceful or destructive purposes. While the artwork is weak and does not always enhance the content, this series is good for the breadth of information it conveys.

64. Energy: A Matter of Choices. 16mm, 22 min., color. Encyclopaedia Britannica Educational Corp., 1973. $290. Gr. 7–12.

Working on the supposition that it takes a lot of energy to run an industrial society, this film suggests that we have enough fuel to meet energy needs for anywhere from 30 to 300 years, then raises the question, why brownouts? A good, but brief history reviews how electricity came to assume such importance in American life, particularly as it became cheaper and more available in the 1930s and 1940s and as power companies pushed increased use to keep the unit cost low. Environmental problems and the costs of remedying them are examined as are new solutions to the energy problem. Conservation and the necessity for government to develop and implement an energy plan are suggested as solutions.

65. Energy: A Study of Resources (Series). 6 filmstrips w/6 cassettes or 6 discs, 58–71 fr., 9–11 min., color, w/tchr's. guide. Q-Ed Productions, 1974. $122.50 ser. w/cassettes; $110.50 ser. w/discs; $12 ea. filmstrip; $8 ea. cassette; $6 ea. disc; $3 tchr's. guide. Gr. 7–12.

The major theme of this series is that our energy crisis presents us with some major decisions to be made about our lifestyle, our choice of new energy sources, and our relations with other countries, and that there will have to be definite environmental and human trade-offs involved. *The Price of Progress* contains good historical frames describing the changes in American life and in energy consumption following industrialization, our dependence on irreplaceable fuels, and our insatiable appetite for fuel for generating electricity and for transportation. *Oil: A Finite Fuel* shows how we are using oil 50,000 times faster than it is made in nature and notes that at current usage rates even the vast Prudhoe Bay field contains enough oil for only two years. It takes a quick look at the problems of oil exploration and the waste involved in our use of oil. The realities of oil shale as an alternate energy source and of environmental and lifestyle changes are also mentioned. *Coal, Natural Gas and Environment* asks if the destruction of the environment is worth the mining of huge coal reserves. Discussing but not describing the Four Corners power plant, it explains the disruption of Indian society engendered by the plant. Natural gas and coal gasification as dependable fuel sources are very briefly discussed and rejected. *Atomic Energy* briefly explains the advantages of nuclear power and the process of generating electricity from it, and mentions attendant dangers from thermal pollution and nuclear accidents. It discusses the breeder reactor and touches on the possibilities of fusion power. The diagrams in this part are not thoroughly labelled. *Future Alternatives* makes an interesting point: America is a space-age society relying on fossil-age fuels. It examines, without enthusiasm, the prospects for solar, geothermal, wind, and water energy as alternate sources; sees some hope in methane and increased power plant efficiency; and asks where we should put our research time and dollars. But it does not provide students with enough information to form opinions. *Politics of Energy* contrasts an American narrator voicing our energy problems and a woman in a have-not nation expressing the desires of underdeveloped nations for energy and its benefits. Noting that energy brings economic and political power, but that most powerful industrial nations fuel their growth with energy from other countries, the filmstrip sees energy supplies used as political tools unless have and have-not nations can share. Each unit is a short overview and does not go into depth on any one topic. Good background material precedes each part but will be limited to the teacher and senior high readers because of the reading difficulty.

66. Energy for the Future. 1 silent filmstrip, 42 fr., color, w/tchr's. guide. Visual Education Consultants, 1974. $10.50. Gr. 7-12.

A quick overview of alternate energy sources, this strip touches on breeder reactors, laser beams, low-polluting gas or synthetic oil from coal, wastes which produce methanol or alcohol, direct or indirect use of solar energy, wind power, ocean temperature differences and tides, and geothermal power. It concludes that there may be limits on how much energy we can generate without badly damaging the environment but does not elaborate. There is good additional information in the teacher's guide, but it must either be read aloud or put on a cassette. The producer has granted permission to do the latter.

67. Energy for the Future. 16mm, 17 min., color, w/tchr's. guide. Encyclopaedia Britannica Educational Corp., 1974. $220; $14/1-3 days rental. Gr. 7-12.

Produced in cooperation with the American Geological Institute, this is a comprehensive, well-edited overview of the alternate energy sources available to this country. The film looks at research facilities for coal gasification; oil shale—showing a laboratory simulation of oil retorting; and the problems surrounding the use of the atom, including the possibility that uranium ore needed to get U235 will be very scarce by 2000. The problems of the breeder reactor are touched on, as are the difficulties of using deuterium to produce fusion energy. Both wet and dry geothermal heat are discussed and the potential for use in the western U.S. noted. Wind research at Oklahoma State University which may overcome the major problem of wind power—vacillating wind strength—and solar energy research and its potential as a supplement to conventional energy in most parts of the U.S. are demonstrated. Includes a look at the Zomeworks model community in New Mexico, where elegantly simple technology uses wind and solar energy exclusively. It may be an eye-opener to many students.

68. Energy: Less Is More. 16mm, 18 min., color, w/tchr's. guide. Churchill Films, 1973. $250; $21/3 days rental. Gr. 7-12.

Made in cooperation with the Environmental Quality Laboratory of the California Institute of Technology, this film emphasizes that we *must* develop new energy patterns to cope with the growing crisis. Four main areas of energy use are discussed: transportation, buildings, appliances, and waste. Automobile mileage has quadrupled in 30 years, plane use has increased 15 times, yet only in rare places—such as the San Francisco Bay Area Rapid Transit System—are efforts being made to discourage urban sprawl and encourage means of transportation more efficient than the car. Ways of building which would combat waste in both cooling and heating are shown. A "solar system approach" which requires a completely new orientation in building design if solar energy is to be used economically is described. The need for laws for high-efficiency appliances and for a complete turn-around from a throw-away society to one which recycles and reuses is emphasized. Former Interior Secretary Stewart Udall asks that we readjust our bigger, better, faster philosophy to the sensible one of "less is more."

69. Energy: New Sources. 16mm, 20 min., color, w/tchr's. guide. Churchill Films, 1974. $280; $21/3 days rental. Gr. 7–12.

This nontechnical film provides a brief survey of several alternative energy sources including energy generated from methane, tidal power, ocean temperature differences, trash, and wind, but its major emphasis is on geothermal power, fusion, and solar energy. The potential for using geothermal power in areas other than the Far West is mentioned, as is the fact that producing energy from any source will have an effect on the environment. Dr. Richard Post, physicist at the Lawrence Livermore Laboratory, suggests a crash program for achieving practical fusion power by the year 2000. The economic and technical problems involved in harnessing solar energy are brought out: the potential use of solar cells for converting the sun's rays directly into electricity if research on mass production of these cells is initiated; the increasing use of solar water heaters for home use as the cost of fossil fuels rises; the possibility of nationwide use of solar power if solar conversion systems can be made insensitive to the weather. The film takes the stance that the government has ignored alternative energy sources, with the exception of nuclear fission, and it must now channel sufficient funds and effort into research on these alternative sources to give consumers more options.

70. Energy: Sources and Man's Needs. 1 filmstrip w/cassette, 65 fr., color, w/25 worksheets and tchr's. guide. Ward's Natural Science Establishment, 1974. $29. Gr. 7–12.

This semi-programmed study unit, intended for individual use, is divided into nine sections with worksheet questions following each section. It provides an overview of energy sources; our consumption patterns; the role of fossil fuels, water, geothermal, nuclear, and solar power; and the environmental effects of some energy sources. Although the content of each section is minimal, the tables and diagrams are good. There is a large amount of audio information for the number of frames. A good format for individualizing energy study, this should lead into materials of greater depth and visual content.

71. Energy: The Dilemma. 16mm, 20 min., color, w/tchr's. guide. Churchill Films, 1973. $280; $21/3 days rental. Gr. 7–12.

Suggested as a social studies film, this could also be used to introduce basic energy consumption and development problems to general science classes. Citing the fantastic demand for fossil-fuel energy which began with the Industrial Revolution and predicting that in three decades U.S. demand will triple, the film concentrates on ways of providing more fuel for electricity and for transportation. The idea of "exporting the pollution" by relocating power plants in the desert hundreds of miles from consumers is rejected when the environmental impact on the desert is shown. Nuclear shortcomings and the environmental and political problems of continuing to bank on oil are pointed out, e.g., a child born in the 1970s will live to see almost all our oil and natural gas used up. The potential of coal and shale oil as sources is briefly examined, but former Interior Secretary Stewart Udall stresses that water is the limiting factor in developing both these resources. Udall takes the stand that conservation is the *only* solution to the dilemma. The film's open-end asks if Americans are willing to find better ways to design buildings, reuse and recycle many of our goods, and develop more efficient transportation by limiting dependence on the automobile—in effect, to make major social changes to alleviate the crisis while alternate sources are developed.

72. Energy to Burn. 16mm, 20 min., color, w/tchr's. guide. BFA Educational Media, 1973. $300; $20/3 days rental. Gr. 7–12.

This Biological Sciences Curriculum Studies film is loosely organized. Its particular strengths lie in showing how energy consumption has increased in staggering proportion to human progress and how a tremendous amount of energy is used to produce a familiar product—a record shown lying in a garbage dump. The amount of energy which went into making, packaging, and distributing the record is documented. A cartoon quiz show called "Take the Consequences" is also included in the film but makes no telling point. Useful to draw attention to energy consumption in general science or social science classes.

73. NewsBank Library. Category: Environment. 1 guide, 12 microfiche, and indices (m., q., a. compilations). NewsBank, Inc. $150 plus shipping. Gr. 7–12.

A current-awareness reference service gathering material from 150 newspapers nationwide, NewsBank has 13 major subject categories. The relevant category is Environment, including such subtopics as Energy Resources, Environmental Issues, Land Use and Land Resources, Mines and Mineral Resources, Natural Resources, Nuclear Power and Radioactive Pollution, and Oceans. A 1977 monthly fiche issue included articles on garbage for fuel, geothermal wells, coal in Colorado, compressed air storage to produce electricity at a future date, the potential of methane for fuel, and the popularity of wood and wood stoves in Virginia and Minnesota. Information on the subtopics is grouped together, but subscribers must take the entire environment category. Subscribers receive a guide to the index as a whole which is updated annually, a binder, monthly fiche, two copies of the monthly and quarterly indices, and one annual index.

74. North Dakota Agriculture: Green and Gold. 16mm, 30 min., color. Bill Snyder Films, 1977. Rental Dist.: North Dakota State Film Library. $375; $3/week rental. Gr. 7-12.

The accomplishments of North Dakota farmers and their efforts to increase productivity in the face of a burgeoning world population are commended in this film. It also shows how energy is both produced and consumed by a major U.S. industry—agriculture. Using footage of current agricultural methods and old photos of farming in North Dakota, the film documents the change from a labor-intensive economy to an energy-intensive one. It briefly explains how photosynthesis transforms the sun's energy into food for humans and animals, and the role of ruminating animals in the energy cycle. Consumption of diesel fuel, gasoline, petro-based pesticides, and fertilizers by the agricultural industry is noted, and a number of mammoth agricultural machines are shown. The makers claim, however, that only 3½% of the total energy consumed annually in the U.S. is used by field agriculture—less than jet aircraft use—while food processing and delivery takes three to five times as much. The importance of weed and insect control to maintain productivity is stressed, and energy-saving measures such as treating seed potatoes with systemic pesticides to avoid repeated spraying later are shown. Scenes of the birth of a calf, a rodeo, young North Dakota farmers, and the work of a country agent round out the film for North Dakota audiences but are peripheral to the subject of energy.

75. A Practical Guide to the Energy Crisis (Set). 2 filmstrips w/2 cassettes or 2 discs, 78-85 fr., 10-11 min, color, w/tchr's. guide. Prentice-Hall Media, 1975. $50 set. Gr. 7-12.

Part I begins with the gas shortage and explains why our energy consumption is outstripping our energy production: wasteful buildings, the inherent in-efficiency of electric power, a four-fold increase in the energy needed to produce our food over the past 50 years, the move to the suburbs, and a throw-away society which demands new energy to make ever-new consumer goods. Part II discusses alternative energies but emphasizes the need to cut energy consumption at the consumer level, noting, among other things, that 46,000 light bulbs were removed from the World Trade Center to conserve energy. Narrated by the science editor of the New York *Daily News*, this set gives a good overview of the problem, although some of its statements, e.g., "nuclear generators will be coming on line at the average of one every three weeks," are now questionable.

76. A Thousand Suns. 16mm, 9 min., color. Barr Films, 1974. $150; $15/5 days rental. Gr. 7-12.

Described as an energy ethic film, and a winner of six film awards, this short, involving film puts our use of energy in the broader perspective of our societal values. In 30 years we have used more energy than all our forefathers. Energy has made us well-to-do and freed us from drudgery. The film asks if we are happier now that only 1% of our work is done by physical power, and we have used energy to build a throw-away civilization. What will we do if we gain the use of fusion, described as the "ultimate energy"? It suggests that the finer alternative is to create values which endure—a sense of wonder and respect for ourselves and all living things. Probably most useful in social studies classes discussing the implications of our energy-hungry society and where the future lies. Users should be aware of brief footage on a baby nursing in the "values" portion of the film.

77. "toast." 16mm, 12 min., color, w/tchr's. guide. Bullfrog Films, 1974, 1977 release. $180; $21.50/3 days rental. Gr. 7-12.

Handsome images show that toast is the end product of an energy-intensive process that begins with the exploration for oil. Oil is transported to the refinery, piped to a conversion site where it is converted into fertilizer and spread on the fields to nurture wheat. The wheat is then shipped to the processing mill to be ground into bread, taken to the stores, sold and carried home. In the end, when a piece of toast is burned and tossed into the garbage, the point of energy waste is driven home. Since this film is virtually without narration, it requires both an introduction and a follow-up regarding the energy uses and changes described.

78. The Future. The Universe and I Series. 16mm or videocassette, 16:32 min., color, w/tchr's. guide. Agency for Instructional Television, 1976. $230 16mm; $135 videocassette. Gr. 8-12.

The first third of this film offers possible answers to the question "Where will the energy to power the world of the future come from?" Researchers explain in varying detail energy from laser fusion, magnetohydrodynamics (MHD), extracting energy from ocean waves or capitalizing on the differences in ocean temperatures, and wind energy. There is no reference to the economics of these methods, their practicality for large-scale energy generation, or their nearness to fruition. The rest of the film is peppered with such ideas as trains of cars running on electromagnetic energy, atomic airships, remote blackboards, underground houses, and human colonies in free space powered by solar energy.

79. Associated Press Special Report: The Fuel Crisis (Set). 2 filmstrips w/2 cassettes or 2 discs, 95-99 fr., color, w/tchr's. guide. Prentice-Hall Media, 1974. $50 set. Gr. 9-12.

Through a series of interviews with energy experts, Part I provides insight into the political and financial reasons behind the fuel crisis. The blame is laid upon our high-energy lifestyle, our wastrel approach to natural resources, tax incentives which have caused oil companies to rely on foreign supplies, government reluctance to come to grips with impending fossil fuel shortages by supporting alternative energy research, and the position of environmentalists on offshore drilling and strip mining. It is a refreshingly frank look at the crisis. Part II examines the changes in the life of the English suburbanite and discusses some effects the energy crisis has already had or may have—smaller cars, fewer trips, moves from suburbia back to the city, possibly unemployment. Optimistic in tone, it quotes several experts who feel we *can* reconcile increased energy demands with respect for the environment, but it is not conclusive about a way out of the problem. The filmstrip can open discussion on how the energy crisis could and should alter our lives.

80. Energy (Series). 5 filmstrips w/5 cassettes or discs, 117–122 fr., 15½–22½ min., color, w/tchr's. guide. Concept Media, 1976. $225 ser. Gr. 9–12.

Part I: The Dilemma is a long but very effective presentation of the current energy problem, strengthened by good interviews with former Interior Secretary Stewart Udall. He takes the responsibility for the Santa Barbara oil spill and says that neither strip-mined coal nor shale oil are solutions and that conservation is the only safe and certain way. Montana ranchers and farmers whose land is threatened by demands for coal state their case convincingly. Unfamiliar pictures in the strip set it off from others covering the same topic more briefly.

Part II: The Nuclear Alternative asks "Is nuclear power a clean, safe source of electricity or is it an invitation to disaster?" The experts voice four major concerns: reactor safety, dangers during transportation, possible destructive use of nuclear materials, and problems of waste disposal. Particularly good are diagrams of what would happen in case of a reactor meltdown. Nuclear physicist Edward Teller places faith in locating nuclear plants underground; the ubiquitous Dr. John Gofman says we cannot accept an option based on plutonium and should free our economic and human resources to work on alternatives. Hans Bethe is quoted as saying that nuclear is the only nonfossil power source we can rely on for some time. The strip concludes that nuclear power does involve certain risks but that the risks are statistically small compared with other risks that our society accepts.

Part III: New Sources covers offbeat sources such as shredded trash, animal manure, offshore wind generators, as well as fusion and solar energy. It endorses solar energy, sees geothermal as an important source that the government must develop, and fusion as more promising than fission. The message is that a number of options should be kept open, and possibly combined, to avoid repeating the mistake of dependence upon one new source, as we did with nuclear energy.

Part IV: Less Is More asks for a reassessment of our whole way of life, keying it to energy realities. It presents the need for more efficient modes of transportation, noting that the least efficient ones—cars and planes—are growing rapidly at the expense of the energy-efficient modes. Waste in homes and office buildings, and the often unnecessary use of large appliances, is discussed, as is Oregon's pioneering law banning all non-returnable beer and soft-drink containers to save energy.

Part V: Actually, You Can Live Better! shows Lee Schipper, an energy specialist at the University of California, Berkeley, in his campaign to prove that the public can live well while saving energy. Noting that Sweden's energy use is half ours, he shows how to save energy in four main areas: heating and cooling, transportation, lighting, and running machines. He maintains that it takes more jobs to save energy than to generate it, refuting the fears of those who see economic crisis if energy-saving measures are prescribed. Overall, this series, based on an award-winning film from Churchill, is characterized by well-edited visuals (some of which are repeated in more than one strip) and a fact-filled but imaginative narrative which makes this a film-like filmstrip program. Despite the number of frames, the parts do not drag.

81. Energy (Set). 2 filmstrips w/2 cassettes or 2 discs, 101–103 fr., 19 min. ea., color, w/tchr's. guide. Prentice-Hall Media, 1975. $50 set. Gr. 9–12.

These filmstrips, part of the producer's *Science and Society* series, are suggested as a multidisciplinary program for use with either science or social science groups, but are probably best used in the latter. Part I raises the question "Can the U.S.—and the world—produce enough energy to keep up with the present growth trends and projected demands?" It gives no firm reply, stating that the ultimate answer depends on current and future fuel supplies, efficient use of available fuels, and our willingness to tolerate some environmental change. Nearly 40 frames are devoted to nuclear power and its attendant problems. The audio content is superior to the visuals, which are technically good but contain many long shots of generating facilities. Part II examines possible new energy technologies. The best section contains diagrams of both the fission and fusion processes, and concludes that these are "not likely to solve our energy problems in the near future." Coal conversion, oil shale, tidal and wind power, and solar energy are examined very briefly and ruled out, with methane production from wastes considered more probable. The alternatives are not examined in depth, but this set is useful as a discussion starter. It raises questions such as: "How much longer will we be willing to accept the deterioration of the environment and the rapid depletion of our energy sources in return for a comfortable lifestyle?" or "Can we learn to make energy conservation not only a fact of life on an immediate personal basis but also on a long-term, societal basis?" It notes, however, that there is no form of energy which does not have some effect on the environment.

82. Energy: A World Resources Project. 1 filmstrip w/cassette or disc, 76 fr., color, w/tchr's. guide, ditto masters, 30 student sourcebooks. Xerox Education Publications, 1974. $49. Gr. 9–12.

A think-set asking what form of energy shall we base our future on. The filmstrip discusses oil ebbing, the problems inherent in coal, the fact that the technology for large-scale use of solar energy does not yet exist, the possibilities of geothermal power, and the pros and cons of nuclear energy. Much of it is presented in the form of a dialogue between an economist and an environmentalist, and the visuals are static. The approach is good although each type of energy does not get equal time; viewers may feel nuclear power is receiving a plug. The ditto masters for attitudinal surveys and student sourcebooks which contain short energy articles are thought-provoking.

83. Energy Sources: A Matter of Policy. Energy Sources: A New Beginning Series. ¾" videocassette or 16mm, 29 min., color, w/tchr's. guide. Univ. of Colorado, Educational Media Center, 1975. $230 videocassette; $333 16mm; $2.50 tchr's. guide. $11/2 days 16mm rental; $11/2 days, $20/15 days, $30/30 days videocassette rental. Gr. 9–12.

Although this film, the first in a series, covers the mechanics of producing energy from geothermal, nuclear, natural gas, shale oil, coal, tar sands, solar, and wind power, its overriding concern is to get the public thinking about making an energy policy before industry or government makes one. Somewhat didactic in tone, it nonetheless presents in considerable detail the realities of such seldom-mentioned sources or techniques as tar sands and natural gas stimulation by nuclear fracturing. It also explains briefly the Energy Policy Project of the Ford Foundation and the options this project opened for future energy use in the U.S. Serious issues are clearly presented and could be discussed in general science, earth science, and chemistry classes.

84. Energy Sources: A New Beginning (Series). 8 filmstrips w/8 cassettes, 37–55 fr., color, w/tchr's. guide. Univ. of Colorado, Educational Media Center, 1975. $100 ser.; $15 ea. Gr. 9–12.

Produced under a grant from the U.S. Department of Health, Education and Welfare, this series contains much more detail on a variety of energy resources than other programs. In each strip, a lecturer, using a studio setting, describes the nature of the source, examines that source's potential, and is frank about its ability to meet America's vast energy needs. There is nothing simplistic about the approach taken; students will find parts of it somewhat technical. On the whole, the narrative is stronger than the visuals. Considering the amount of detail presented and the relatively low cost, it is recommended for senior high students who want content rather than generalities.

(I) *The Sleeping Giant: Coal* (41 frames) claims that we have as much coal as the Arab nations have oil. The strip shows the problems entailed in obtaining coal: the hazards and economics of underground mining, the damage to the land from surface mining, the high sulfur content of coal reserves, and the location of these reserves. Desulfurization and coal conversion to more environmentally acceptable fuels are also discussed. The narration is so brisk and fact-filled that frequent pauses may be needed to allow viewers to absorb the material.

(II) *Oil Shale: The Rock that Burns* (41 frames). Although there are oil-shale beds in 34 continental states, shale has never been seriously considered as an energy source because new oil fields are still being discovered. This unit (whose visuals are not on par with the narrative) describes both above-ground retorting and *in situ* retorting. It notes the potential economic and environmental impact of shale production upon the three large shale-reserve states of the Green River Formation and concludes that even with large-scale development, shale will go only a short way toward solving our energy problems.

(III) *Tar Sands: Future Fuel* (37 frames). A misnomer for bitumen, tar sands offer the advantages of being closer to the surface, easier to separate from host rock, and requiring a minimum amount of processing before refining. This strip examines these sands and their disadvantages— principally that 85% of the material is waste when surface mining is performed. Thermal recovery underground has not yet proved to be economically viable. Tar sands are offered not as a permanent solution, but as a source which could provide the time extension needed until Americans find energy technologies which no longer take from the earth.

(IV) *Geothermal Power: The Great Furnace* (45 frames). One of the most difficult in the series, this strip states that the extent, distribution and size of geothermal resources in the U.S. is poorly known. It describes the differences between the vapor or steam system, the water-dominate system, extraction from hot, dry rock, and the binary system which extracts energy from low-temperature geothermal reservoirs which are hot enough to vaporize low-boiling-point gases for energy use. The fact that some terms are not defined but shown only in the diagrams may limit its usefulness. The vocabulary is more technical than in the other strips.

(V) *Solar Power: The Giver of Life* (55 frames). Discusses the four known methods of converting raw solar energy into useful forms of power: low-temperature thermal (heating a liquid or gas through contact with a sun-absorbing surface); high-temperature thermal (parabolic mirrors or a "solar farm"); electric or photovoltaic (the solar cell); and chemical conversion by photosynthesis. The diffuseness and variability of solar power are noted, as are the cost of collectors and the investment required for solar devices. Stating that the essential research has been done, the strip asks when national energy policy will include a serious commitment to the solar area.

(VI) *Wind Power: The Great Revival* (39 frames). Thought to be too expensive until now, wind power may become competitive with the rising cost of conventional fuels. This strip explains the principle of windmills and their potential for both small- and large-scale power generation. It cites the joint efforts of the National Science Foundation and NASA to evaluate wind energy, which have been hamstrung by the lack of funds to do large-scale research on this free, clean, and non-depletable form of energy.

(VII) *Gas Stimulation by Atomic Fracturing* (38 frames). Once considered an unhealthy byproduct of crude oil, gas, according to this strip, has become our prime energy provider for needs other than transportation. Fracturing is done to stimulate production from areas that do not yield readily their natural gas. Despite its title, this strip describes drilling and hydraulic fracturing as well as explosive fracturing of underground gas by conventional or nuclear methods. The producers admit that these tests have not been encouraging: flow rate has declined, gas quality was substandard, and there was a slower decline in radioactivity than anticipated.

(VIII) *Nuclear Energy: The Great Controversy* (43 frames). The title of this unit is somewhat misleading, since the bulk of the strip explains the method of obtaining energy from U-235 and the principle of the breeder reactor. Only the last eight frames focus on the actual controversy. The producers acknowledge the potential danger from plutonium but do not present any other facts of the controversy, except to state flatly that nuclear plants do not pollute the air and that "a very serious problem is the possible release of an enormous amount of radioactivity to the environment." Neither statement is elaborated on.

85. Power and Energy (Set). 2 filmstrips w/2 cassettes or 2 discs, 49–59 fr., 12–15 min., color, w/tchr's. guide. Educational Activities, 1972. $29 set. Gr. 9–12.

Part I explains the limitations on the use of fossil energy and is particularly good in its diagrams and explanation of how electricity is generated from hydroelectric, geothermal, and fossil-fuel sources. Terms like power grids and peak power loads are introduced and the techniques of scrubbing and electrostatic precipitation of pollutants are explained. It also discusses how the fundamental laws of science, such as those involving magnetism and radiation, are put to practical use. Part II describes the structure of the atom and the fission process, leading into an explanation of how an atomic pile is used to produce electricity. It predicts that by the year 2000, 50% of our electricity will come from nuclear sources. It presents alternative sources of energy such as fuel cells, man-made steam pockets, magnetohydrodynamics, and thermoelectric power and describes the technical problems surrounding fusion as a source of unlimited power. The program optimistically concludes that "whichever technologies emerge as the main source of power, you can be sure that they will no longer be a threat to our environment."

86. Energy (Series). 3 filmstrips w/3 cassettes, 103–110 fr., 12–21 min., color, w/tchr's. guide. Hawkhill Associates, 1976. $75 ser. Gr. 10–12.

This series is a moodpiece that synthesizes the scientific and humanistic viewpoints about energy. *Eternal Delight: What Is Energy?* takes the view that we are not running out of human energy and details the energy cycle from the sun to man to man's work and back again. *The World Game: Energy in Human History* traces the history of energy, from the interdependent harmony of man, animals and plants, to man as a slave to energy as the means for survival. It then discusses the change wrought by machines and their inventors, a new kind of slavery, and reminds the viewer that half of the world's population is still the loser in the energy game. *Good Morning, Brother Sun* notes that less than 2/100 of 1% of the 173 billion megawatts of solar energy received by the earth daily is trapped and used. It recounts the history of man's use of energy and reveals that today Americans command 10,000 watts each, over 100 times as much as our primitive ancestors. Discussing the Oregon "energetics" study which concluded that solar energy is the logical alternative and that conservation— the need to do more with less—is a major concern, the filmstrip gives a surface overview of what can be done with solar power to slow down the "restless, ceaseless, one way flow of energy." There is a philosophical addendum in which the mandala, a simple diagram synthesizing all we know about energy and its potentials, is the focus. Mainly for humanities classes, this visually attractive series requires a discussion leader to help get its message across.

87. The New Alchemists. 16mm, 29 min., color. National Film Board of Canada, 1974. Dist.: Benchmark Films. $3.95; $40 rental. Gr. 10-12.

Convinced that man is exploiting unrenewable resources and fouling his environment, a group of young scientists has designed and set up new means of using solar and wind energy to live off of 12 rented acres on Cape Cod. Their efforts are shown too briefly to provide an introduction to either wind or solar energy, but the film does show applications like solar fish ponds and new-design windmills. Use for enrichment in general science or biology classes.

88. The Other Way. NOVA Series. videocassette or 16mm, 26 min., color, w/tchr's. guide. Time-Life Multimedia, 1975. $235 videocassette; $335 16mm; $35/3 days rental. Gr. 10-12.

A provocative film from the NOVA series which asks whether we should be seeking an alternative style of life rather than alternative forms of energy. Fuel economist E. F. Schumacher questions the "20th Century myth that bigger is better." He does not suggest a return to the primitive but rather an intermediate technology in which we are not obsessed with labor-saving machines at the sacrifice of all other factors. His thinking has had a major impact on Third World nations where labor is plentiful but for whom sophisticated machines are expensive and, he suggests, unnecessary, as they compound the unemployment problem. Schumacher also questions the future of cities, which may be unsupportable if a small fraction of the population must be fuel intensive to support them. He questions the logic of eliminating the human factor in manufacturing, believing that creativity and self-satisfaction have been lost in the quest for more efficiency and more leisure time. While nuclear energy has been suggested as the ultimate answer to energy problems, Schumacher notes that to make it a mass phenomenon, society must go to the breeder reactor, producing lethal plutonium. Beyond this, he insists that nuclear is an "energy sink," i.e., we spend more energy installing nuclear plants and policing their wastes than is returned in energy. A refreshing, insightful look at the energy problem set in the framework of our technological society, this production would be ideal for social studies classes discussing 20th Century issues or science classes concerned with the implications of our fuel-intensive society.

89. Which Energy. 16mm, 23 min., color, w/tchr's. guide. Stuart Finley, 1976. $350; $35/1 day, $47.50/1 week rental. Gr. 10-12.

This film recalls the energy shock of the gasoline shortages of 1973-74 and asks where the emphasis should be placed for obtaining energy in the future. It describes, but does not explain, experiments with fusion technology, the current use of solar energy for home heating, existing geothermal uses, experiments with wind power, the unlikely future for oil shale, and the ebbing status of oil and natural gas. The potential of coal and nuclear power receives most consideration. The film concludes that coal is important as the way to give us breathing time to develop other resources, but that reserves are worthless unless real environmental protection methods are developed. The nuclear footage concentrates on the possibilities of the breeder reactor. It shows experiments at Hanford aimed at insuring safety, but notes that the first commercial breeder was closed down when a partial meltdown occurred. Advocating conservation and a necessary concentration on several energies, this film presupposes knowledge of energy terminology and some science interest.

Coal

PRINT

90. National Coal Association. **Coal in Today's World.** Washington, D.C.: National Coal Assn., 1974. 24pp. free. Gr. 3-7.

This attractive booklet explains how coal is formed, how it is used, types of mines and mining, where coal can be found, and the history of coal in America. It does not discuss environmental impact, except for pointing out the need to reclaim lands which have been surface-mined. New terms are defined. A lesson plan centering on this booklet is available.

91. Chaffin, Lillie D. **Coal: Energy and Crisis.** Irvington-on-Hudson, N.Y.: Harvey House, 1974. 47pp. $3.99 plb. Gr. 4-6.

The author, who grew up in the Kentucky coal fields, writes authoritatively about mining methods, the unstable business of coal mining, and its increased mechanization. She points out facts about the many environmental problems which accompany mining and examines coal's place in the energy future. A concise and informative book.

92. Ridpath, Ian, ed. **Man and Materials: Coal.** Reading, Mass.: Addison-Wesley, 1975. 32pp. $4.95 plb. Gr. 4-6.

The emphasis in this well-illustrated, informative book is on the mining of coal rather than its energy potential. Self-contained spreads discuss how coal was formed, the history and distribution of coal, and mining methods over the centuries. Transportation and utilization of coal and coal by-products are also considered. It includes good, well-labelled diagrams for science reports.

93. Burt, Olive W. **Black Sunshine: The Story of Coal.** New York: Messner, 1977. 63pp. $6.64 plb. Gr. 4-7.

Believing that many children today have never seen coal, the author describes how coal is formed, used, and mined. The narrative revolves around the life of a Pennsylvania coal miner who takes the reader through an underground mine shaft. The dangers involved in underground and strip mining are discussed, as are improvements in mine safety and the life of the miners. A glossary is included.

94. Kraft, Betsy Harvey. **Coal: A First Book.** New York: Watts, 1976. 66pp. $4.90 plb. Gr. 5-7.

This generally thorough book describes the formation of coal, the history of its use, changes in its mining and distribution, how it is mined, the hazards of mining, and the life of a miner yesterday and today. The use of coal as an important source of electric power and its inherent environmental problems are discussed, as are efforts to clean up air pollution, to reclaim mined land, and to turn coal into less polluting fuels. Includes many photographs.

95. Doty, Roy. **Where Are You Going with That Coal?** Chicago Museum of Science and Industry Series. New York: Doubleday, 1977. 62pp. $5.95. Gr. 5-9.

This work provides a brief look at coal's history, its use in generating electricity and in making iron and steel, and the phenomenal range of coal by-products. The sections on coal gasification and South Africa's SASOL complex, which turns coal into synthetic petroleum, are particularly interesting. The prospects for coal conversion as a solution to the energy crisis are discussed. Appealing in format, this book is enhanced by Doty's clever, informative two-page cartoon illustrations of each topic. A glossary is included.

96. National Coal Association. **What Everyone Should Know about Coal Gasification.** Washington, D.C.: National Coal Assn., 1977. 15pp. free. Gr. 7-9.

A scriptographic booklet which explains how coal can be turned into nonpolluting gas. This booklet uses cartoons and brief questions and answers to convey information, but students using it will need some understanding of chemistry.

97. Greenbaum, Margaret Elaine. **Kentucky Coal Reserves: Effects on Coal Industry Structure and Output.** Lexington, Ky.: ORES Publications, 1975. 23pp. $4 pap.; $1.50 microfiche. Gr. 10-12.

This 1975 study provides a survey of changes in the structure of the coal industry from 1965 to 1972. It notes that coal production during this period increased much more rapidly in Kentucky than it did nationally, primarily because of surface mining in Kentucky, but as more lands have been surface-mined, productivity per mine has declined. Future growth of the coal industry in Kentucky will require increasing numbers of smaller-scale mines. Greenbaum is not optimistic about industry expansion of underground mining, since surface mining has considerable cost advantages. Useful in local or regional studies of coal's potential as an energy source.

98. Schweitzer, Stuart A. **The Limits to Kentucky Coal Output: A Short-Term Analysis.** Lexington, Ky.: ORES Publications, 1974. 25pp. $3 pap.; $1.50 microfiche. Gr. 10–12.

In this brief, somewhat technical 1974 study, the author—in the Department of Economics at the University of Kentucky—is not optimistic about the ability of the coal industry in Kentucky to meet the rapidly growing needs for coal, particularly with regard to underground mining. He notes, however, that surface mining tonnage could increase by 20–25%. The two variables in growth potential appear to be the ability of the industry to employ its equipment and labor more efficiently and the willingness of the industry to do so. Useful for students in that state or region.

99. Toole, K. Ross. **The Rape of the Great Plains: Northwest America, Cattle and Coal.** Boston, Mass.: Atlantic Monthly Pr., 1976. 271pp. $8.95. Gr. 10–12.

Toole, an historian and a Montana rancher, argues fervently against the hurried exploitation of Montana's coal reserves in the name of "Project Independence." Carefully documenting the actions of the coal and power companies, he writes of the legal, sociological, and ecological objections to strip-mining and mine-mouth power generation to produce electricity for the Pacific Northwest. He describes the combatants in the battle to make Montana a "national sacrifice area" and provides two interesting chapters on Indian reaction to the pressure of energy interests. Standing behind the ranchers, environmentalists, and other Montanans opposed to coal exploitation, Toole deplores plans for the massive appropriation of the state's water resources. He questions whether the fragile plains land can ever really be reclaimed and suggests that plains coal has been oversold as a low-sulfur energy reserve in the interest of quick profits.

NONPRINT

100. Earth Resources: Coal. Discovery Series. 1 audiocassette, 12 min. Associated Educational Materials, 1975. $6.50. Gr. 4–8.

A clear, well-paced introduction to coal, its history and its uses. The narrator notes that despite decreased use of coal for home-heating and railroads, coal continues to be essential for the steel industry and for generating electric power. He describes how coal is formed and traces the gradual acceptance of coal as a power source in the U.S.: the Franklin stove and the railroads made coal popular, the factory and the machine age made it indispensable. At the conclusion, three questions based on the content of the tape are asked. The cassette provides good basic information for reports being prepared by the upper elementary or middle school students, especially those who have reading problems.

101. Mining for Coal. wall chart. 18" x 14", color & b&w. National Coal Assn., n.d. free. Gr. 4–8.

This two-color poster explains in black and white photographs how coal is mined and the equipment used in mining.

102. Using Coal/Moving Coal. wall chart. 18" x 14", color & b&w. National Coal Assn., n.d. free. Gr. 4–8.

Two of the ways coal is used—in utility plants and coke ovens—as well as ways in which it is transported are explained in this chart.

103. Coal: The Rock That Burns. 16mm, 13 min., color, w/tchr's. guide. Centron Educational Films, 1976. $205. Gr. 5–9.

A somewhat glossy paean to the coal industry, this describes our reserves, defines four kinds of coal, and gives some inkling of the hard, dirty work which underground mining still entails despite mammoth machines and safety techniques. It notes that half of today's coal supply comes from surface mining. While allowing that poor or inconsiderate mining practices left land devastated in some areas, the film claims that today's mine operators respect the environment and provide for public recreation areas and grading and replanting of land which has been strip mined. It acknowledges that sulfur dioxide is a pollutant associated with burning coal, but states that pollution controls are expensive and difficult to install. Coal by-products such as coke, coal tar, light oil, etc. are mentioned.

104. Coal, Corn and Cows. 1 filmstrip w/cassette. 96 fr., 18½ min., color, w/tchr's. guide. Hawkhill Associates, 1977. $24.50. Gr. 7–12.

This program describes the dilemma of central Illinois, a rich farming area which is blessed with abundant coal. It describes how the coal and topsoil were formed and notes that high agricultural productivity is due in part to fossil fuels. The question is: can farms and coal mining coexist or must the standards and way of life of the area change? There are no pat answers here, but it is a good discussion starter on a problem made more real by the localization of the issue.

105. Stripmining in Appalachia. 16mm, 25 min., b&w. Appalshop Films, 1973. $250; $35 rental. Gr. 7–12.

A low-key, somewhat rambling overview of what strip mining has done to Appalachia. Considerable footage is included of scarred or polluted land and earth-moving equipment and shovels gulping as much as four tons of earth in a bite to get at the coal. Interviews are shown with a coal company representative who states that "after it's stripped, the land is much better property," an old man whose home and surroundings have been ruined, and an engineer who has seen areas he reforested in the early days of strip mining stripped again thirty years later (he feels strip mining should never have been allowed). The film's predictions, that an eightfold increase in energy demand in the next forty years will result in a threefold increase in the demand for coal, and that if strip mining continues to grow as it has, an area nearly as large as Massachusetts, Connecticut, Rhode Island, and Delaware will have been stripped by the year 2010, seem quite plausible, if appalling.

106. The Sleeping Giant—Coal. Energy Sources: A New Beginning Series. ¾" videocassette or 16mm, 29 min., color, w/tchr's. guide. Univ. of Colorado, Educational Media Center, 1975. $230 videocassette; $333 16mm; $2.50 tchr's. guide; $10/3 days 16mm rental; $10/3 days, $20/15 days, $40/30 days videocassette rental. Gr. 9–12.

We have more coal than the Arab nations have oil, yet coal in 1973 supplied only 17% of our country's energy because of its inherent problems: bulk, the hazards of underground mining, the damage from strip mining, the pollutants it releases when burned. This film, somewhat static but fact-filled, supplies answers to the question "Where does coal fit into our energy future?" The film states that coal production must double by 1985 for the U.S. to achieve energy self-sufficiency and acknowledges the time, money, and work force needed to develop new underground mines and the environmental problems of strip mining. Although there is some footage of strip mining and reclamation in Colorado, the bulk of the film is studio-based, using wall charts and rear-projection to show the problems and possibilities that coal presents. The narrator discusses the likelihood of converting coal by means of gasification and methanation to useful, clean alternate fuels; these processes are charted but not demonstrated. A comprehensive, structured film which is somewhat technical in the last few minutes, it mentions all the problems inherent in coal production but does not go deeply into any of them.

Electric Power

PRINT

107. Wade, Harlan, and Wrigley, Denis. **Electricity.** Milwaukee, Wisc.: Raintree Children's Books, 1977. 32pp. $4.49 plb. Gr. K–2.

This simple picture book explains where electricity in a lamp comes from— beginning with the power plant. It defines the terms "electric current," "conductor," and "insulator." The uncomplicated cartoon-in-a-frame design helps to get the points across.

108. Grey, Jerry. **The Race for Electric Power.** Philadelphia, Penn.: Westminster, 1972. 125pp. $5.95. Gr. 7–10.

Although the sections on fission and fusion are now somewhat dated, the chapter on how electric power is generated and distributed and the chapter on generating sources such as the breeder reactor and magnetohydrodynamics provide good basic information with diagrams. There is some discussion of pollution problems.

109. Edison Electric Institute. **31 Answers to 32 Questions about the Electric Utility Industry.** New York: Edison Electric Institute, 1976. 36pp. single copies free; bulk prices range from $.33 to $.75. Gr. 9-12.

Using data derived from the *Statistical Year Book of the Electric Utility Industry* and from Federal Power Commission publications, this regularly updated booklet answers questions about the use of electricity at home and by industry, the number of kilowatt-hours generated by major fuels, nuclear plants in operation, and the economics of the electric-utility industry in clear, readable form. A glossary is included. This may also be available directly from local investor-owned utility companies.

110. Hamer, John. "The Future of Utilities." **Editorial Research Report,** Vol. 1, No. 10. Washington, D.C.: Editorial Research Reports, 1975. 18pp. $2.50. Gr. 9-12.

The three sections of this state-of-the-industry report cover electric utility trends, pressures on the regulatory structure, and conflicting approaches to utility reform. The report provides a good wrap-up of the problems besetting the industry, consumer reaction, and how these two factors affect industry expansion to meet projected energy needs. Students will find it easy to read. The bibliography includes books, articles, studies, and reports.

111. National Electric Reliability Council. **Fossil and Nuclear Fuel for Electric Utility Generation: Requirements and Constraints, 1977-1986.** Princeton, N.J.: National Electric Reliability Council, 1977. 26pp. free. Gr. 10-12.

NERC's members include all of the electric-power systems in the U.S. and four in Canada. This short publication takes up the requirements of the U.S. electric industry for fuels, the availability of those fuels, and the constraints on the use of those fuels which are anticipated for the next decade. It projects increasing dependence on coal and nuclear fuels, less on gas and hydroelectric power, with oil contributing about the same percentage. Presenting the utility-industry position on energy development, it states that the industry's ability to assure adequate electricity is threatened by federal surface mining legislation, items in the National Energy Plan and the current Administration's "ambivalence toward nuclear power, together with the hiatus in commitment toward the breeder reactor for longer-term needs."

112. Hackleman, Michael A. **Electric Vehicles: Design and Build Your Own.** Mariposa, Calif.: Earthmind/Peace Pr., 1977. 202pp. $7.95 pap. Gr. 11-12.

This is a good guide which covers both the theory behind and practical instructions for construction of an electric vehicle, as well as the legal problems of putting an electric vehicle on the road. Students using it will need advanced knowledge of auto mechanics and electricity.

NONPRINT

113. A Powerful Friend (Series). 4 filmstrips w/1 cassette, 3½–8 min., color, w/tchr's. guide and 4 activity sheets. Prod.: Consumers Power Co. of Michigan, 1975. Rev. by Edison Electric Institute. Dist.: Spartan Graphics. $20 ser. Gr. K–1.

Originally produced by Consumers Power Company of Michigan and revised by the Edison Electric Institute, this program is now available from the vendor or from local utility companies. The distributor recommends that it be presented in nine steps, with four activity sheets alternating with each of the four filmstrips. The song *Power* is first introduced as a sing-along and presents the idea of power as "the thing that makes things go." Filmstrip I, *Power and Energy*, begins a cartoon dialogue between a boy and Franklin, the talking mule, who explains that machines have replaced him on the farm and that animals, people, and machines all have power coming from energy. The concept of energy is not well explained. In *How Do We Get Power*, Franklin and the boy talk about the ways electricity is used in homes, schools, and factories and learn that generators make electricity. The third filmstrip discusses air pollution and the ways factories can control it, defines environment and the need to reclaim land which has been mined for iron and coal, and presents the idea of recycling while plugging electricity as the way to clean the air and run recycling plants. The final strip, *How You Can Help Use Energy Wisely*, uses a blown fuse in the boy's home to introduce the idea of "peak" and "off-peak" periods for electricity use and suggests that children find out when the peak period is in their area so they and their families can use electricity during off-peak hours, thereby reducing the need to build additional power plants. Since there is so little material available for this age group and the price is relatively low, this production is endorsed as an introduction to electricity and its use. Elementary teachers felt that it should *not* be used alone but should be part of a general lesson on energy, and that it should definitely be introduced and followed up with class discussion. The strips themselves could be effectively used by individual children or small groups. Though produced for grades K-3, these teachers did not feel the content merited use above the first grade. Worksheets are adequate, but individual teachers may wish to prepare their own materials with more emphasis on new words and their meanings.

114. Lightning to Light Bulbs. 1 filmstrip w/cassette, 30 fr., 15 min., color, w/tchr's. guide. Encore Visual Education, 1972. $17; $9 filmstrip only; $8 cassette only; $1 tchr's. guide. Gr. 1-5.

Using lightning as a starting point, this strip offers a very basic introduction to what electricity is, how electricity gets to the home, and the functions of the transformer, the circuit breaker or fuse, and the light switch. Visuals are sketchy but adequate. The reverse side of the cassette, called *The Unseen Worker,* is a conversation between Jack, the child who received the explanation on side one, and an anthropomorphic "Electricity," who gives a casual explanation of the terms AC-DC, voltage, ampere, and watt. This is supposed to motivate Jack to investigate further on his own.

115. Electricity: Changing and Transferring Energy. 1 filmstrip w/cassette or disc, 54 fr., 9:22 min., color, w/tchr's. guide. Society for Visual Education, 1976. $18.50. Gr. 5-9.

The purposes of this strip are to stimulate students' interest in learning about electricity; to define energy, its sources, and its forms in everyday life; and to define electricity and identify it as a form of energy which can be produced from and transferred to other forms of energy. It provides a good presentation of atomic structure, the basic principles of electricity, and how one form of energy is changed into another. The narration is somewhat condescending, and comments about energy needs at the end are an appendage rather than an integral part of the unit.

116. Building an Electric Generator. 1 silent filmstrip, 43 fr., color. Encyclopaedia Britannica Educational Corp., 1967. $10. Gr. 7-12.

This strip helps students answer the following questions: How does a generator produce electricity? What is an alternating current? What is a direct current? It graphically demonstrates the principle that when a coil is moved through an electric field, it generates electricity by converting the mechanical energy (the up and down motion) into electrical energy (the electric current). A simple generator is constructed and alternating and direct currents are explained through diagrams and experimentation. This strip should be useful where there is an emphasis on developing reading skills in subject areas.

Energy Conservation

PRINT

117. Smaridge, Norah. **Only Silly People Waste**. Drawings by Mary Carrithers. Nashville, Tenn.: Abingdon, 1976. 32pp. $4.25. Gr. K-4.

This collection of more than a dozen funny rhymes with wild and woolly illustrations depicts some of the more familiar ways in which kids waste resources and explains why they shouldn't waste. The first four rhymes focus specifically on energy: turning off lights, closing the refrigerator door, turning off the television, and keeping sticky fingers off the thermostat. The author makes her point.

118. Epstein, Sam, and Epstein, Beryl. **Saving Electricity.** Illus. by Jeanne Bendick. Champaign, Ill.: Garrard, 1977. 64pp. $4.48 plb. Gr. 3-5.

When a blackout hits the town of Riverton, Jimmy Morris and his class get lessons about electric power generation, energy consumption, and alternate sources of energy. Jimmy concludes that since alternative energy sources cannot meet large-scale needs yet, the only solution is to save energy at home. Jimmy gets his family involved in the effort to conserve energy. The advantages of recycling paper, cans, and bottles to save electricity then used to make new products are also mentioned. This is meaty reading for the middle grades.

119. Thomas Alva Edison Foundation. **Energy Conservation Experiments You Can Do.** rev. ed. Southfield, Mich.: Thomas Alva Edison Foundation, 1977. 32pp. $21/100. Gr. 4–7.

This booklet exhorts students to become energy-waste watchers and offers 11 experiments centering on energy consumption and waste around the average home. The final section, which includes instructions for building a sun-powered hot-dog cooker, could use more and better illustrations.

120. Electrical Industries Association of Southern California. **The Energy Savings Guide Book.** Los Angeles, Calif.: Electrical Industries Assn., 1973. 13pp. $.25. Gr. 4–12.

This guide provides tips for reducing day-to-day energy costs at home and in the car. Suggestions are grouped under such topics as "Comfort Conditioning," "Water Heating," and "Laundry Equipment." Each section includes several humorous cartoons to attract interest.

121. Marshall, James. **Going, Going, Gone? The Waste of Our Energy Resources.** New Conservation Series. New York: Coward, McCann, 1976. 94pp. $5.49 plb. Gr. 6–12.

The author believes that despite the 1974 gasoline shortage, Americans continue to waste huge amounts of fuel. He discusses where we get our energy and at what cost to the environment, and the potential of nuclear power and other energy sources. The emphasis of this book is on the need for individuals, industries, and the federal government to use energy more intelligently and develop conservation practices and policies in keeping with the diminishing energy supply. A better-than-average YA title, this work does not hedge on the issues and maintains reader interest while documenting the problem.

122. Energy Conservation Research. **Our Energy—Problems and Solutions.** Malvern, Penn.: Energy Conservation Research, 1977. 48pp. $1; discount available on orders for 50 or more copies. Gr. 7–12.

This compact, two-color guidebook explains what energy is, its forms, energy conservation laws, and how we use energy. Its major emphasis is on recommendations for energy conservation in running automobiles, heating and cooling, and a variety of household and agricultural operations. The section "Energy for the Future" does not present a complete picture of the pros and cons of each energy source, and the page of "additional sources" primarily lists trade associations representing segments of the energy industry. This may be available from local public utilities at little or no cost.

123. Felton, Vi Bradley. **150 Ways to Save Energy and Money.** New York: Pilot Books, 1977. 40pp. $2.50 pap. Gr. 7–12.

This easy-to-read handbook focuses on inexpensive ways to deal with rising energy costs and dwindling energy resources. It describes how to save energy in lighting and in heating and cooling a home, how to use appliances economically, how to save hot water, and, for apartment dwellers, how to get neighbors involved in conserving energy. One chapter gives practical advice on how to cut down on gas guzzling.

124. Sunbeam Appliance Company. **Making Less Electricity Do More.** Oak Brook, Ill.: Sunbeam Appliance Co., n.d. 15pp. free to teachers; limited quantity for classroom use at no charge. Gr. 7–12.

Contending that small appliances such as electric frypans consume less energy than electric ranges or ovens, this booklet compares the use of both for several cooking purposes, lists electric appliance operating costs, and offers 13 recipes which can be used in cooking with small appliances.

125. United Nations Environment Programme. Office of the Executive Director. **The State of the World Environment, 1978.** New York: United Nations Environment Programme, 1978. 12pp. free. Gr. 7–12.

The final portion of this report on contemporary issues of international significance considers energy conservation. It notes that more than half of the energy put into daily use is wasted because of losses induced by technology and by man. The emphasis is on obtaining more work from fuel which is already being consumed. Brief, specific suggestions for improvements in housing and lighting and in the agricultural, transport, and industrial sectors are made; effective conservation measures in both developed and developing countries are mentioned. No detail is given but this is a clear, thoughtful summary.

126. Cornell Cooperative Extension Energy Task Force. **Save Energy Save Dollars.** Ithaca, N.Y.: New York State College of Human Ecology and New York State College of Agriculture and Life Sciences, 1977. 95pp. $1.50 postpaid; 75 or more copies, $1.10 ea. Gr. 9–12.

This well-written manual is intended for consumers who are trying to conserve energy resources and to reduce expenditures for energy and fuel consumption. Most topics are oriented towards home owners, e.g., "Hot Water: Use and Conservation" and "Be Energy Conscious When You Buy a House," but these sections and the three-page "Energy Checklist" at the back of the book could also be helpful in home economics, consumer education, and occupational education classes. One section is specifically titled "Energy Conservation for Young People." The book also includes several useful tables such as "Types of Storm Windows and Doors," "Characteristics of Northeastern U.S. Woods for Fireplace Use," and "Estimated Annual Kilowatt Hour Consumption for Selected Electric Appliances."

127. Friend, Gil, and Morris, David. **Kilowatt Counter: A Consumer's Guide to Energy Concepts, Quantities and Uses.** Milaca, Minn.: Alternative Sources of Energy, 1975. 36pp. $2. Gr. 9–12.

Presented as a "tool for making informed decisions regarding our individual buying habits, energy consumption and conservation, and environmental responsibility," this publication from *Alternative Sources of Energy* magazine intends to do for the national energy diet what calorie counters do for people. Its usefulness lies in the charts and tables showing energy consumption by appliances and modes of passenger transport, fuel efficiencies for heating and water heating, and solar insolation received in 28 North American locations.

128. How You Can Save Energy Every Day. Greenfield, Mass.: Channing L. Bete, 1977. 15pp. $.50; discount on bulk orders. Gr. 9–12.

This is a scriptographic booklet which combines words and drawings to show how to save energy inside and outside the house, in the yard and workshop, and when driving and shopping. It could be used effectively in consumer education classes with students who are not motivated to read.

129. Darmstadter, Joel. **Conserving Energy: Prospects and Opportunities in the New York Region.** Baltimore, Md.: Johns Hopkins Univ. Pr., 1975. 105pp. $8.95; $2.95 pap. Gr. 10–12.

Although energy demand in the New York City region has grown at a rate less than in the U.S. as a whole, New York's dependence on liquid fuels makes it extremely vulnerable to energy shortages. Darmstadter, senior research associate at Resources for the Future, a Ford Foundation research group, documents projected energy consumption for the area without conservation measures and with "dampened growth." He concludes that even modest energy-conserving practices can postpone expected 1985 levels of consumption by five years. More adequate home insulation, space heating and cooling, and improved automotive fuel efficiency could signal major savings. He cautions that regional conservation efforts are more likely to succeed if undertaken within a national program, since this would eliminate or reduce "border-crossing" to evade restrictive measures. He notes that well-intentioned solutions to the conservation problem are often too simplistic, ignoring practical trade-offs. A thoughtful, well-reasoned report, documented in tables and text, this will be limited to bright students because of style and vocabulary.

130. Hayes, Denis. **Energy: The Case for Conservation.** Worldwatch Paper No. 4. Washington, D.C.: Worldwatch Institute, 1976. 78pp. $2 pap. Gr. 10–12.

Taking the stance that "in the immediate future, saved energy is our most promising energy source," this paper explains how energy savings can be accomplished in the areas of transportation, buildings, food, and waste products. Hayes distinguishes between curtailment and conservation and also tackles the controversial area of the relationship between energy, the Gross National Product, and labor. He includes tables on energy consumption and efficiency, and recommends six ways individuals can conserve energy. This is a rational but highly readable paper with enough detail to satisfy the layman and the student preparing a research paper. It is adapted from Hayes's book *Rays of Hope: The Transition to a Post-Petroleum World* (see No. 175).

131. Morrell, William H. **The Energy Miser's Manual.** Eliot, Me.: Grist Mill, 1976. 77pp. $1.95 pap. plus $.30 postage. Gr. 10–12.

Miserliness is back in. The author gives hands-on advice about insulating and weather-stripping houses, and he explains why these processes are necessary. He provides useful statistics for school reports on how to save energy and money. Tables on appliance operating costs and saving which can be realized from modifying a "characteristic" home, and chapters on house heating, fireplaces, and gasoline savings, round out this environmentally oriented, generally nontechnical guide.

132. Morris, David, and Friend, Gil. **Energy, Agriculture and Neighborhood Food Systems.** Washington, D.C.: Institute for Local Self Reliance, 1975. 16pp. $.75 pap. Gr. 10–12.

The authors state that our agricultural system relies on fossil fuels and that food processing alone accounts for over one-third of the energy devoted to food in the U.S. They advocate rejection of energy-intensive, expensive chemical fertilizers derived from fossil fuels in favor of natural fertilizers and urban gardening.

133. New York State College of Agriculture and Life Sciences. **Energy Facts.** Factsheets Series. Ithaca, N.Y.: New York State College of Agriculture and Life Sciences, Cornell University, n.d. 2–4pp. ea. free. Gr. 10–12.

Each of these 20 illustrated factsheets takes up an important aspect of consumer energy conservation. Titles include: "The Value of Storm Windows and Doors," "Water Heating: Use and Conservation," "Energy Efficiency in Major Appliances," "Energy Efficiency at Mealtime," and "Your Energy Efficient Car." Although one in the series, "Energy Conservation for Young People," is particularly directed to ages 12–16, most of these sheets would be useful in senior high home economics or consumer education classes.

134. Smith, Thomas W., and Jenkins, John. **The Household Energy Game.** Madison, Wisc.: Univ. of Wisconsin Sea Grant College Program, 1974. 20pp. single copies free; $.15 ea. on multiple orders. Gr. 10–12.

This booklet uses a point and grid system to help consumers account for their personal consumption of energy. It asks consumers to compare their energy consumption level to that of the average American household in 1955 and today, as well as to energy consumption levels in other countries. The second half of the booklet is devoted to ways consumers can modify their energy "budget" to save energy and money.

135. Derven, Ronald, and Nichols, Carol. **How to Cut Your Energy Bills.** Farmington, Mich.: Structures Pub. Co., 1976. 131pp. $4.95 pap. Gr. 11-12.

Besides offering advice on weather barriers, ventilation, lighting appliances, and hot water heating, this how-to book includes a chapter on heating and cooling which explains the operation and advantages of the heat pump, a well-illustrated chapter on energy-saving homes, and fall and spring checklists for conserving energy at home.

136. Eccli, Eugene, ed. **Low-Cost, Energy-Efficient Shelter.** Emmaus, Penn.: Rodale Pr., 1976. 408pp. $10.95; $5.95 pap. Gr. 11-12.

Specific chapters of this easy-to-read how-to title discuss insulation, savings on heating and cooling equipment, saving money with appliances and lights, water heating and water conservation, designing the solar-tempered home, the greenhouse as a source of food and winter heat, and solar water heaters. Written by architects, engineers, and scientists, the book's primary use will be in heating and air-conditioning courses.

137. Leckie, Jim, et al. **Other Homes and Garbage: Designs for Self-Sufficient Living.** San Francisco, Calif.: Sierra Club Books, 1975. 302pp. $9.95 pap. Gr. 11-12.

Somewhat cryptically titled, this paperback is an outgrowth of Stanford University's Workshop on Social and Political Issues. It represents an attempt by eco-oriented engineers to communicate practical technical information about energy-efficient living to nonspecialists. Portions on climatic factors in house design and on solar heating which include discussions on thermal comfort, climatology, heat loss in dwellings, and heating systems will be of interest to advanced students in heating and air-conditioning or architecture programs. The section on small-scale generation of electricity from wind or water is geared to more advanced students in electrical programs who have had intermediate algebra. Other parts of the book deal with waste-handling systems, water supply, and agriculture/aquaculture.

138. Price, Billy L., and Price, James T. **Homeowner's Guide to Saving Energy.** Blue Ridge Summit, Penn.: Tab Books, 1976. 288pp. $8.95; $5.95 pap. Gr. 11-12.

Primarily aimed at the do-it-yourself market, this detailed but plainly written book offers good advice on insulation, weather-stripping, heating and cooling systems, appliances, lighting, and wiring for the vocational or consumer education student or any student who is destined to become a homeowner or renter.

NONPRINT

139. Saving Energy. 1 poster, 24½" x 32", color, Abt Publications, 1977. $2.50 plus $.50 for postage and handling. Gr. K-4.

A bright, four-color, four-seasons cutaway of the Average American Home demonstrates 15 easy ways to conserve energy around the house. Among them are: caulking windows, putting insulation in the attic, setting the air conditioner lower, having an expert check the furnace's efficiency, keeping the stove and refrigerator well apart, etc. Boldly done on 110-lb. paper, this poster will brighten the primary grade classroom or children's room.

140. Saving Energy Kit. 1 Saving Energy game and poster, 1 copy of "In the Bank or Up the Chimney," 1 copy of "Tips for Energy Savers," 14 activity sheets, w/tchr's. guide. Abt Publications, 1977. $19.50. Gr. 3-7.

The goal of this package is to raise the consciousness of intermediate and upper elementary students, to make them aware of the energy conservation problem and more knowledgeable about its solutions. The basis of this kit is a brightly colored Saving Energy board game which can be played by up to five students. With each role of the die, they learn from "Energy Pointer" cards and, later, "Letter Question" cards which build on the knowledge a-massed earlier. The player with the most energy points wins. Purchasers should be aware that the "gameboard" is not a board but a poster which the producers advise be covered with plastic; the same poster without the overprinted game numbers sells for $2.50 separately. The replicable student activity sheets can be used by the entire class and vary in intended grade level. The simplest is a maze drawing puzzle for grade three; the most complex, "Thermometer Thinking," is a math word problem related to energy conservation and requires knowledge of percentages and fairly complex multiplication. These worksheets are creative and closely related to the experience of elementary children. The two reference pamphlets provide the teacher with energy-conservation background. The teacher's guide provides instructional objectives, consciousness-raising questions, the rules of the game, and ideas for follow-up discussion and homework assignments.

141. Joey's World. ¾" videocassette or 16mm, 23 min., color. Stephen Bosustow Productions, 1974. $300; $15/1 day, $7.50 ea. additional day rental. Gr. 3-12.

A moodpiece about what kind of world seven-year-old Joey, the narrator's son, will have a quarter century from now. Joey is shown playing, unaware of man's appetite for raw materials, while the narrator asks where Joey will find enough energy and raw materials to maintain the life Americans are accustomed to. The film claims that our energy needs have increased enormously but that the earth hasn't "cooperated" by providing additional energy sources. A man and a woman on the street express the idea that we must return to a more simple life and that the technology which created our energy problem will now have to rescue us. There is a dramatic cut to open sewers pouring out America's refuse—4.5 billion tons of garbage a year. The film's final point is that we must conserve, reuse, recycle, and weigh use against environmental destruction. Its somewhat naive conclusion—that Joey will find a way to meet the energy needs of his generation—weakens the film's conservation message.

142. Conservation Through Recycling: From Air Pollutant to Product, Pt.I—Scrap Iron and Steel. 1 filmstrip w/1 cassette, 120 fr., 18 min., color, w/tchr's. guide. Current Affairs, 1971. $22. Gr. 7–10.

Somewhat tangential to the energy problem, this program requires teacher direction to indicate the links between energy conservation and recycling. Impressive statistics are noted: America produces one billion tons of solid waste annually, and as many as 20 million cars or a billion dollars worth of raw materials in cars are going to waste. It notes that using a ton of iron scrap to make new steel would save a ton of coke, a half ton of limestone, and one-and-a-half tons of iron ore.

143. Energy and You: Conserving Energy. 1 silent filmstrip, 42 fr., b&w, w/tchr's. guide. Visual Education Consultants, 1973. $7.50. Gr. 7–12.

The classic methods of conservation (small cars, public transportation, good driving habits, proper auto maintenance, car pooling, home insulation, restricted use of air conditioners) as well as some lesser known ones (such as proper-fitting refrigerator doors and switch-operated electric starters for gas appliances) are all simply illustrated. The teacher's guide provides good additional material with which to understand each frame; the filmstrip's usefulness is limited without the narration. Teachers unwilling to read it in a darkened room should note that the producer permits the content to be put on a cassette.

144. How to Save Energy. 1 filmstrip w/1 cassette, 76 fr., 10 min., color, w/tchr's. guide. Encore Visual Education, 1974. $21. Gr. 7–12.

After showing a mind-boggling array of appliances, this program gets down to the nuts and bolts of how to read an electric meter, what a kilowatt-hour is, how much energy is used by different electrical appliances, and practical, specific ways of conserving energy around the house, from running a full washing machine at night to putting the right-size pot on the right-size burner. Good for general science or home economics classes teaching home management.

145. Living with a Limit: Practical Ideas for Energy Conservation. 160 slides w/2 cassettes and 2 discs, 30 min., color, w/tchr's. guide. Science and Mankind, 1978. $139.50 ser.; $86.44 ea. pt. Gr. 7–12.

Part I begins with President Carter's statement that we waste more energy than we import and proceeds to offer specific suggestions for conserving energy for schools and for automobile drivers. The slides demonstrate suggestions for schools such as cooler rooms, later openings, longer vacations during high fuel consumption months, fewer lights in certain areas, and stepped-up maintenance work to aid in conservation. The emphasis in the driving section is on careful maintenance and driving and fewer trips in more economical cars. The slides in this portion do not always demonstrate clearly the points made by the narrative. Part II explodes the "my home is my castle" myth with regard to energy waste. The narrator mentions ways of preventing heat loss in winter. Concurrently, a series of cryptic slides show an icicle-bedecked house; illustrating the points literally might have been more effective. This same problem occurs in the series of visuals on how to cook and use the refrigerator efficiently—some slides again are too general to illustrate the well-made points. Sections on choosing an air conditioner and conserving hot water are tied together better; a section on alternate sources of energy for America seems tacked on at the end. A good quality series, but it could have been better edited.

146. Saving Energy on the Road. videocassette or 16mm, 15min., color, w/instruction sheet and discussion questions. Ramsgate Films, 1976. $225; $20/3 days 16mm rental. Gr. 9–12.

A tight, to-the-point film illustrating four main ways to reduce driving costs with the particular objective of conserving gasoline: choosing the smallest, lightest car possible; using practical, gas-saving techniques such as gradual acceleration and braking and maintaining moderate, steady speed; observing proper maintenance by techniques such as setting the choke lean, checking the air filter and changing oil regularly, and removing any unnecessary weight from the car; driving less, by coordinating shopping, using public transportation, car pooling, and rediscovering walking, hiking, and biking. Because of its length, good acting, and several light notes—one of the people in the car pool is seen shaving in the car—this film should be well received in all driver education classes and in social studies classes considering practical ways to save energy.

147. Energy Management. 16mm, 15 min., color. Walter J. Klein, Ltd., 1975. $275; 24 hr. free loan. Gr. 10–12.

This attractive, neatly produced film takes the position that windmills and water wheels can't supply the energy needs of industrial society, and fossil fuels are not unlimited, so we must husband our resources by adapting management techniques to energy use. Several dozen at-home techniques are shown, from cooking more than one dish at a time in the oven to installing time controls for programming heat, lighting, and ventilation systems. Some time is spent on how building design can be part of wise energy utilization. With the exception of one statement that "coal and refuse are always plentiful" as energy sources, the film appears accurate and useful for home economics or general science classes studying the energy problem.

148. Focus on Energy Conservation. ¾" videocassette, 29 min., color. Public Issues Network, 1977. $85; $7/day rental. Gr. 11–12.

Part of the Focus series produced by the Carnegie Endowment for International Peace and the Brookings Institution, this panel discussion brings together three experts in the field of energy. They discuss conservation, trends such as low energy prices, the growth of oil imports, and feasible patterns for energy conservation and consumption in America. The role of the government in setting standards for energy-efficient buildings and the pros and cons of price regulation, at both production and consumption levels, are discussed. This tape touches on the economic aspects of the energy problem and has some interesting things to say about U.S. energy patterns compared to those in industrialized nations in western Europe.

Energy—Environmental Aspects

PRINT

149. Concern, Inc. **Eco-Tips #7.** 4 parts. Washington, D.C.: Concern, Inc., 1974. 2–8pp. ea. Part 1, $.30 or $13.50/100; Part 2, $.35 or $16/100; Part 3, $.35 or $16/100; Part 4, $.20 or $8.50/100. Gr. 6–12.

Four brief folders provide state-of-the-resource information on fossil fuels (part one), nuclear energy (part two), solar energy (part three), and geo-thermal energy (part four), while emphasizing the environmental or health hazards of each source. The publisher is a nonprofit, tax-exempt organization whose aim is "to develop public awareness of environmental issues and to recommend appropriate citizen action." An Energy Packet including these four as well as *Eco-Tips #5*, "Energy Conservation," and booklets on Federal energy resources and new energy technologies for buildings is available for $2 including postage.

150. McCoy, J. J. **A Sea of Troubles**. Greenwich, Conn.: Clarion/Seabury, 1975. 184pp. $7.95. Gr. 6–12.

The second chapter, "The Fouling of the Sea," of this well-researched book discusses pollution by the growing number of oil spills, by tritium from nuclear reactor operations and fuel processing, and by heat from both nuclear and nonnuclear sources.

151. Adamson, Wendy Wriston. **Who Owns a River? A Story of Environmental Action.** Minneapolis, Minn.: Dillon Pr., 1977. 96pp. $5.95 plb. Gr. 7–12.

In 1965, the Northern States Power Company announced plans to build a 550,000 kilowatt power plant on the St. Croix River, bounded by Minnesota and Wisconsin. Although citizens' groups objected to the thermal, air, and visual pollution which was bound to result, the plant was built. The fight demonstrated the need for federal water pollution control legislation and better regional planning and resulted in the National Environmental Policy Act, the Federal Water Pollution Control Act Amendments of 1972, and the Wild and Scenic Rivers Act. Using many photographs, Adamson provides historical and ecological background of the St. Croix area while raising a basic energy issue: which has priority, power or the preservation of the environment?

152. Center for Science in the Public Interest. **99 Ways to a Simple Lifestyle.** Bloomington, Ind.: Indiana Univ. Pr., 1977. 381pp. $12.50; $3.50 pap. Gr. 7–12.

A whole-earth approach to living on what Adlai Stevenson called "our fragile craft." Based on a 1975 Harris poll which found that 77% of the American population are willing to simplify their way of life, this book offers 99 specific changes which Americans can make to accomplish that goal.

The chapters "Heating and Cooling," "Solid Waste," and "Transportation" are energy-related, as are portions of the action-oriented concluding chapter "Community." Sources of additional information are given after each of the 99 suggestions.

153. Hellman, Hal. **Energy in the World of the Future.** New York: Evans, 1973. 240pp. $6.95. Gr. 8–12.

How we can produce more energy and do less damage to the environment is the focus of this well-researched title in the World of the Future series. After a prologue in which he envisions what life might be like in the next century, Hellman takes a comprehensive look at fossil fuels, fission, and the energies of tomorrow. There are good sections on air- and other types of pollution and on energy storage. Some of the data and the predictions are now dated.

154. Wallace, Daniel, ed. **Energy We Can Live With; Approaches to Energy That Are Easy on the Earth and Its People.** Emmaus, Penn.: Rodale Pr., 1976. 150pp. $3.95. Gr. 9–12.

A brief, recent, and highly personalized potpourri of articles from Rodale's publications *Organic Gardening, Compost Science,* and *Environment Action Bulletin.* This collection encompasses pedal power in the Orient; the outlook for methane and windmills; food policies that save money; the economics of a wood cookstove; and a solar heat-collecting pond, to name but a few. As the title indicates, these are approaches rather than detailed how-to articles.

155. Brown, Theodore L. **Energy and the Environment**. Columbus, Ohio: Charles E. Merrill, 1971. 141pp. $3.50 pap. Gr.11–12.

This is an important book because it goes into some of the environmental effects of energy use in real detail. The author describes, for example, the effects of atmospheric carbon dioxide and particulate matter and the localized consequences of heat generation on bodies of water. He contends that "the continued burning of fossil fuels at the rates now projected into the next century would result in a very sizable increase in the average global temperature, with disastrous consequences in terms of climatic change," and he recommends a national energy policy including the establishment of controls on the rates of growth of various components of the energy market. The information is presented in a semi-scholarly, didactic fashion which will limit its use to better students with considerable interest in the topic.

156. Energy and Power. A Scientific American Book. San Francisco, Calif.: W. H. Freeman, 1971. 144pp. $5 pap. Gr. 11–12.

The articles in this book were first published in the September, 1971, issue of *Scientific American.* They place energy consumption in a long-term perspective. The flow of energy in the biosphere and in hunting, agricultural, and industrial societies is considered. In the concluding chapter, "Decision

Making in the Production of Power: How Does a Society Reconcile the Need for Energy and the Finiteness of the Earth?," Harvard Law School professor Milton Katz describes the legal framework within which society can act on energy decisions. This book is for good science students who are also aware of social and environmental responsibilities.

NONPRINT

157. Energy and Our Environment (Series). 4 filmstrips w/4 cassettes or 2 discs, 48–55 fr., 9–12:14 min., color, w/tchr's. guide. Coronet Instructional Media, 1975. $65 ser. Gr. 5–8.

The series addresses itself to two modern problems: energy and pollution. The human use of energy is presented from prehistoric times to today in the first strip, *Man and the World's Energy*, which shows how pollution increased as man's habits and fuel choices changed. *Waste and Pollution* centers on the waste heat and energy which accompany man's use of engines and how the heat and chemical exhaust fumes damage our physical environment. It also dispels the myth that electrical energy is pollution free. *Our Growing Use of Energy* links the energy shortage and the pollution problem to our mechanized way of life and to the increasing population. It presents a dazzling rundown of statistics relating to energy consumption and foresees no change in the energy situation until the world population stabilizes. In *The Future*, the possibilities of reduced energy consumption and alternate energy sources are examined and the crucial philosophical questions are asked, e.g., are we willing to accept a slower pace of life and to sacrifice conveniences in order to clean up the environment and conserve enough energy for tomorrow? For social studies or science use.

158. Energy and the Environment: How to Live on the Earth a Long Time without Using It All Up (Series). 4 filmstrips w/4 cassettes or 4 discs, 96–107 fr., 13–18 min., color, w/tchr's. guide. Macmillan Library Service, 1974. $90 ser. Gr. 5–10.

Where Energy Comes from and Where It Goes is a basic but thorough explanation of various kinds of energy, the sun as the start of it all, and uses of energy in everyday life. *Energy and Its Cost to Our Environment* points out that energy is not only used directly, e.g., by cars, but also to manufacture cars and the things associated with them. It graphically shows the effects of pollutants emanating from energy consumption upon plants, things, and people, and the effects of the searches for new sources of energy, such as oil shale, upon the land. *Harnessing New Sources of Energy* presents a good overview of alternative sources, including tree farms and garbage burning. Particularly good is its discussion of the problems surrounding nuclear fission and fusion. *Energy, Environment and the Future* examines problems concerning the ways we now consume energy—habits established when energy was cheap and plentiful—and the implications for the environment if we continue consuming at our present rate. It asks: Is more and more, with its attendant problems, what we want for the country? While the content is excellent, the visuals lose impact because they are washed out. Ambience is provided by ecological guitar music such as "Whose Garden Was This?"

159. The Dent's Run Project: Rebirth of the Land. 16mm, 27 min., color. Consolidation Coal Co., 1976. $150; free loan. Gr. 6–12.

This film shows a demonstration project, run by Consolidation Coal Company with help from the Federal Environmental Protection Agency and the West Virginia Department of Natural Resources, to reclaim a 14-square mile area in northern West Virginia which had been gutted and polluted by coal mining over the years. The goal was to bring the land back to its original state and to treat acid-laden, tomato-colored water from both active and abandoned mines so that the water would again be usable. Touting the project as an example of cooperation between private industry and government, the film describes and shows the land being bulldozed, contoured, graded, and seeded but is not specific about the land's current utility. It concentrates on the three million gallons of water being diverted and treated each day. Using before and after shots and imposing scenics, it states that "man works with nature to control the changes we make in our world," but it does not get down to hard facts about the success of this reclamation or what effects it has had on the people near Dent's Run. Interesting both as a public relations film and for its recognition of the pollutant problems of mining.

160. Human Issues in Science, Unit II: Energy (Series). 4 filmstrips w/4 cassettes or 4 discs, 79–96 fr., color, w/tchr's. guide. Scholastic, 1975. $69.50 ser. Gr. 6–12.

An issues-oriented series which is divided into four parts. *Energy and the Land. . . ?* goes directly into the environmental impact of strip mining, the human toll in underground mining, and the effects of high dams for hydroelectric power. The goal is to enable the student to weigh America's dramatic energy needs against damage done by energy-making activities. *Energy and the Sea. . . ?* pulls no punches about the effect of oil spills and the damage done by regular industrial and mechanical wastes. It claims that pollution is on the increase. *Using Energy. . . ?* provides a staggering array of energy-consumption statistics with Dantesque examples of air, water, and visual pollution against the song "Whose Garden Was This?" It concludes that the technological wonderland has backfired and we are running out of fuel, and asks what decisions for the future must be made. *The Future of Energy. . . ?,* using pop-art illustrations and hyper-dramatic narration, envisions what energy-poor life could be like in 2024. In nontechnical language, it sets forth both sides of the nuclear power dispute and then focuses on ways of utilizing solar energy. Each section raises a decision-making question regarding our energy needs vs. the effects on people and the environment. Although it will probably be most used in social studies classes, it could also serve as a thinkpiece for general or even laboratory science classes.

161. The Buffalo Creek Flood: An Act of Man. 16mm, 40 min., color. Appalshop, 1973. $200; $20/1 day rental. Gr. 7–12.

In February, 1972, a giant coal-waste dam collapsed in Logan County, West Virginia, leaving 125 dead and 4000 homeless. The Pittston Coal Company, which had built the dam, called the disaster "an act of God." Investigations by the Bureau of Mines, Bureau of Reclamation, and the Geological Survey agreed that the dam was not meant to handle the amount of rainwater which had fallen. The damage was estimated at $50 million, but Pittston, the largest independent coal company in the U.S., settled out of court for $13 million. This film documents the aftermath of the disaster—the losses, the inability of people to recreate their normal lives, and the record of the coal company and government agencies with regard to violations, fatalities, and accidents. It is a strong indictment of the coal industry's attitude that "you keep on producing coal until something goes wrong" and makes a moving case for honest people being disregarded and even sacrificed to keep the coal moving and the profits high.

162. Oil Spill! 20 slides, color, w/written narration. Educational Images, 1977. $18.50. Gr. 7-12.

Twenty well-chosen, good-quality slides depict the results of a modest oil spill—only 308,000 gallons—near Alexandria Bay in the St. Lawrence River in June, 1976. The efforts to clean it up by means of floating booms, mechanical skimmers, pumps, and hand labor are shown, as are the tragic effects the spill had on wildlife: muskrats, fish, ducks were suffocated, geese drenched. The effects on homes, marinas, and property are discussed but not shown. Total cost of the clean-up operation to the federal government was $8.5 million. The long-range ecological effects on the area are only speculated. Recommended because of its careful documentation, brevity, and price.

163. Oil! Spoil! Patterns in Pollution. 16mm, 16 min., color. Sierra Club, 1972. Dist: Association Films. $275; $7.50 rental. Gr. 7-12.

A handsome, nonnarrated Sierra Club film which relates the patterns in America's insatiable demand for petroleum products to the famous Santa Barbara oil spill. The film moves from the lovely California seashore to Sergeant Sweeney's gas station with its recorded greetings, an auto graveyard, a decorated oil pump rising and falling more and more quickly, and the day-and-night stretches of automobiles for which the state is famous. Finally we get an aerial view of the oil spill, which tarred the beaches and required a massive cleanup. A visual lesson on the ecological effects of our obsession with oil, this may need an introduction for students to whom the spill is ancient history.

164. Critical Issues in Science and Society: The Great Energy Debate. 159 slides w/2 cassettes or 2 discs, 33 min., color, w/tchr's. guide. Science and Mankind, 1976. $124.50. Gr. 8-12.

This two-part program casts energy problems as values issues. It examines our hunger for electricity and raises the ethical issues implicit in exploiting new areas of fuel supply. Using case studies such as the environmental effects of the Four Corners generating plant in the Southwest and the long-range impact of strip mining valuable Western farm and grazing land, it confronts viewers with the issues inherent in making energy decisions. Part II opens with brilliant pictures of wrecked cars among beautiful and unbeautiful landscapes, fixing the viewer's attention on the link between the automobile and our energy problems and noting that, because of the car, the U.S. consumes twice as much energy as countries with comparably high living standards. It briefly considers other energy possibilities and the need to decide where to put research dollars; it concludes that energy use carries a price tag and Americans must decide what is worth the price and what is not. Each part of this set contains two discussion breaks based on the case studies given. The slides used do not duplicate visuals seen in other sets and are of exceptional quality. Because of its thrust, the main use will be in social studies classes discussing contemporary problems.

165. Energy vs. Ecology . . . The Great Debate. 16mm, 28 min., color. Consolidation Coal Co., 1974. $150; free loan. Gr. 9–12.

Despite its title, this film does not convey the sense of a great debate. It begins with a TV-commercial look at an area of eastern Ohio in which the Hanna Coal Company, a Consolidation subsidiary, is reclaiming for pasture land which has been surface-mined. It briefly recounts our energy consumption history and advocates coal as our best future energy source. Two main side effects of mining are brought up: the degradation of the land and air pollution. This film insists that mined land can be reclaimed and that in the Great Plains area where our main surface coal reserves exist, the land is actually improved in reclamation. Admitting that sulfur, one of the main pollutants resulting from burned coal, presents a more difficult problem, the film asserts that sulfur *can* be removed after burning and that coal gasification offers a solution to the energy problem. Useful for contemporary issues classes as an example of the industry side of the ecology debate rather than as a balanced picture.

166. Northern Plains Coal. 138 slides w/1 cassette, color. student version, 160 slides w/1 cassette. Matson's, 1973. $96; $25/7 days rental. Gr. 9–12.

A concerned, well-documented appraisal of what widespread coal strip mining will mean to the Northern Plains states in terms of air and water pollution, the destruction of good ranch and farm land, and of the society in this area. As a portent of things to come, the Four Corners power plant in New Mexico is shown in a setting heavy with air pollution—34 tons of particulate and 170 tons of sulfur dioxide a day despite anti-pollution devices. The question is raised as to whether the same fate awaits the Northern Plains. The producer makes a strong case that water will be polluted or diverted for industrial power generation. Feelingly presented, this set is sincere, not slick. It brings home what will probably happen when insatiable power demands from outside an area make mining that area profitable and asks the question: Are the benefits really worth the cost? The student version has intermittent 30–45 second pauses in order to allow questions on two accompanying worksheets to be answered.

167. A Question of Values. 16mm, 28 min., color. The New Film Co., 1972. $350; $35/1 day, $10/ea. additional day rental. Gr. 9–12.

Drawn against lovely coastal scenery, this finely edited documentary probes the public and private debate over placing an oil refinery in the deepwater of Penobscot Bay, Maine. The area is poor, with high unemployment and deteriorating housing, and the refinery would create jobs. The oil company president, a longshoreman whose living has been precarious, and a real estate agent see its positive side, but the newspaper editor, the lobsterman, and the owner of the boat yard believe that if the refinery comes, marine life and life for people will never be the same. While teenagers debate the pros and cons of the refinery, the Maine environmental agency asks if it is good for the state, and a citizen at the public hearings states what he thinks would happen: "The coast of Maine as we know it will be gone and it will be gone forever." Ed. note: The Maine State Environmental Improvement Agency voted against the proposal to build because of the impact of the proposed refinery on the environment.

Energy—Political and Economic Aspects

PRINT

168. Archer, Jules. **Legacy of the Desert: Understanding the Arabs.** Boston Mass.: Little, 1976. 214pp. $6.95. Gr. 6–9.

Chapters 13 and 14 of this young adult title briefly outline the development of "The Arab Superweapon—Oil" in the Middle East and the effects that oil riches have had on that area. The remainder of the book explores Arab history and culture and traces the history of the Arabs' dispute with Israel.

169. Congressional Quarterly. **Continuing Energy Crisis in America.** Washington, D.C.: Congressional Quarterly, 1975. 124pp. $5.25 pap. Gr. 9–12.

This collection of short, self-contained articles covers such varied topics as energy self-sufficiency, Project Independence, Arab oil money, energy sharing with Canada, deregulation of natural gas, offshore oil drilling, and oil company political contributions. The appendix recounts the creation of the Energy Research and Development Administration, federal solar and geothermal energy programs, and the abortive year-round daylight-savings time plan. Useful as historical background on attempts to cope with the energy crisis.

170. Congressional Quarterly. **The Middle East: U.S. Policy, Israel, Oil, and the Arabs.** 3rd ed. Washington, D.C.: Congressional Quarterly, 1977. 196pp. $5.25 pap. Gr. 9–12.

Nearly 60% of the proved oil reserves of the world are in the Middle East. This work offers up-to-date profiles of Middle Eastern nations and analyzes the major issues affecting Middle Eastern political and economic life. The chapter "Middle East Oil" examines how oil has been used as a political weapon during the past five years and describes U.S. energy policy, including Carter's National Energy Plan. Tables on oil prices, production, and imports point out the growing dependence of the Western nations upon the Organization of Petroleum Exporting Countries (OPEC).

171. Grossman, Richard, and Daneker, Gail. **Jobs and Energy.** Washington, D.C.: Environmentalists for Full Employment, 1977. 22pp. $2 to individuals; $5 to institutions. Gr. 9–12.

This well-documented booklet refutes the oft-repeated statement that only by stepping up large-scale energy development can unemployment be contained. The authors contend that energy, once available, generally ends up replacing jobs, noting that the major employment increases since 1947 have been in the low-energy merchandising and service areas of the job market. Numerous examples are offered to support their belief that conservation measures are a desirable alternative to energy expansion and that solar energy development has been neglected. The authors conclude that there will be more jobs if the nation cuts back on energy waste and moves vigorously towards solar energy.

172. Commoner, Barry. **The Poverty of Power: Energy and the Economic Crisis.** New York: Knopf, 1976. 314pp. $10; Bantam, $2.75 pap. Gr. 10–12.

Commoner is the maverick scientist who has spearheaded the "science information movement" which seeks to bring scientific facts to the public attention. In this work Commoner asserts that the three great problems of today—the energy crisis, environmental pollution, and the economic crisis—are inextricably linked and that the root of the three problems lies in our economic system. After discussing thermodynamics, the current state of fossil fuels, and nuclear and solar power, he concludes that the energy crisis was spawned by the industry's profit-motivated decisions. He advocates the development of a production system which is consciously intended to serve societal needs, rather than maximizing private profits. This is a controversial book and Commoner has been criticized for venturing outside his field of expertise. The book reads well and could be the basis for debates in advanced social studies classes.

173. Ford Foundation Energy Policy Project. **A Time To Choose: America's Energy Future.** Cambridge, Mass.: Ballinger, 1974. 511pp. $16.50; $7.95 pap. Gr. 10-12.

The Energy Policy Project of the Ford Foundation has constructed three different scenarios of possible energy futures for the United States through the year 2000. The first scenario, "Historical Growth," foresees a shift to coal and nuclear power as a means of continuing energy use at the 1950-1970 growth rate; it posits that environmental chances must be taken with both sources to produce an enlarged energy supply. The second, "Technical Fix," represents a conscious national effort to use energy more efficiently through engineering know-how, thereby cutting the energy growth rate. The third, "Zero Energy Growth," assumes a redirection of economic growth away from energy-intensive industry towards economic activities requiring less energy, primarily by a decreasing emphasis on making things and an increasing emphasis on offering services. "ZEG" would allow smaller-scale energy sources to share in the total energy demand. One chapter describes the energy patterns of Americans of all income levels and concludes that slower energy growth would *not* have a disastrous effect on employment (since the industries which are the most energy-intensive employ relatively few people). Other chapters take up the following topics: U.S. energy policy in the world context; energy and the environment; private enterprise and the public interest; reforming electric utility regulation; federal energy resources; and energy research and development. The thoroughness with which these topics are treated varies. This study is well documented, written in plain English, and for the most part provides the kind of hard data lacking in many other energy books. Although there is no index, the self-contained chapters can be successfully used for reference purposes. This work is essential for anyone examining the economic and social ramifications of energy development for the future.

174. Foreign Policy Association. **Great Decisions '78.** Boston, Mass.: Allyn and Bacon, 1978. 96pp. $2.73 ea. in quantity. Gr. 10-12.

Each year Great Decisions selects eight major foreign policy topics and presents comprehensive articles on them for use by secondary students. In the 1978 edition, the article "Dilemmas of World Energy: Options for a Fuel-Guzzling Superpower" puts U.S. energy problems in an international framework. The article provides the facts about and raises questions concerning: our addiction to oil; production vs. conservation as solutions in the next two decades; OPEC as a spur to energy reform; and alternate energy sources, particularly nuclear energy. Student opinion on three big energy issues is solicited on a ballot which is bound with the issue. Suggested discussion topics and readings are listed. This is a thoughtful, up-to-date overview of energy choices facing the U.S. The writing style and format will limit it to the good to advanced reader.

175. Hayes, Denis. **Rays of Hope: The Transition to a Post-Petroleum World.** A Worldwatch Institute book. New York: Norton, 1977. 240pp. $9.95; $3.95 pap. Gr. 10-12.

Hayes, an expert in the field of energy research, believes that our generation must take the responsibility for planning and making the transition to a "post-petroleum world" and that we must act swiftly to build an energy economy based on renewable resources. He examines the future for fossil fuels, the problems inherent in nuclear power, conservation as an important immediate energy source, and the outlook for safe, sustainable sources. Implicit in his suggestions are major social changes, e.g., eliminating dependence upon the automobile, decentralizing industry to utilize diffuse fuels. He offers no worldwide panacea, suggesting instead that each nation or area choose the energy strategy that best meets its needs. A thoughtful, well-written book which places the energy problem in world perspective and offers a positive base for making practical decisions. There is good discussion and research material here for the intermediate to advanced social studies student.

176. Morgan, Richard; Riesenberg, Tom; and Troutman, Michael. **Taking Charge: A New Look at Public Power.** Washington, D.C.: Environmental Action Foundation, 1976. 100pp. $3.50 pap.; discount available on orders for 10 or more copies. Gr. 10-12.

Written on the premise that "public ownership of electric utilities may provide a solution to the private power companies' never ending cycle of higher rates, more power plants, and greater environmental destruction," this paperback measures public power against its privately owned competition and suggests methods of organizing a switchover to public ownership. Appendices include a list of "Hydroelectric Dams Subject to Recapture" and an outline of federal statutes and court decisions which the authors maintain demonstrate a legal basis for public ownership of electric utilities. The audience will be social studies teachers and students who are oriented toward citizen action.

177. Ridgeway, James. **The Last Play: The Struggle to Monopolize the World's Energy Resources.** New York: New American Library, 1973. 373pp. $1.95 pap. Gr. 10-12.

Reporting in 1973, Ridgeway contends that the energy crisis has been orchestrated by major energy companies working in concert with government policies or lack of them. He documents the mergers and cartels which have led to a new "energy industry"—interlocks between oil, coal, gas, and uranium owners—and goes beyond U.S. borders to describe the development of energy resources in Canada and in the "Southeast Asia co-prosperity sphere." A long section gives profiles of major energy companies, concentrating on what they own and who owns them. Ridgeway recommends citizen action to block the sale and lease of resources in the federal and state domain and to replace "the private government of energy with truly public institutions" such as regional energy agencies which would develop binding economic plans for their areas. This title provides background for the student who is interested in the economic maneuvering behind our energy problems.

178. Committee for Economic Development, Research and Policy Committee. **Key Elements of National Energy Strategy.** New York: Committee for Economic Development, 1977. 32pp. $1. Gr. 11-12.

Taking a free-enterprise approach to the formulation of American energy policy, this pamphlet comes out strong for promoting increased supplies of presently usable fuels, increasing coal and nuclear capacity while accelerating private research on alternate technologies, and above all, relying on the market system—with minimal government intervention—as the mechanism for stimulating both energy conservation and new energy supplies. For advanced social studies and economics students.

179. "Energy and Employment." **Council on Economic Priorities Newsletter,** Dec. 28, 1977. New York: Council on Economic Priorities, 1977. 4pp. $1; inquire for multiple copy prices. Gr. 11-12.

This issue provides provocative background to the theory that there is a direct relationship between growth in energy consumption and a healthy economy. It describes CEP's Jobs Study, scheduled for release in mid-1978, which will quantify the regional employment and economic effects of two proposed Suffolk County, New York, energy systems: one, additional power plants, and two, a system combining energy conservation and implementation of solar energy. The study aims to determine the actual effect of building additional generating facilities upon the employment rate, and to help local planners in choosing their own energy system by assessing the cost, labor, and materials of each option. Written clearly in a style accessible to average to good readers, this newsletter issue would be helpful in classes in which the economic implications of energy-related decisions are being examined.

180. Freeman, S. David. **Energy: The New Era.** New York: Walker, 1974. 386pp. $14.50. Vintage, $2.45 pap. Gr. 11-12.

Freeman, former director of the Ford Foundation's Energy Policy Project and head of the President's Energy Policy staff, offers a comprehensive look at the energy crisis as of 1974. In nontechnical style addressed to the general reader, he sets up the problems and explores pathways toward the solution and toward a national energy policy. Of particular interest are sections on the conflicting maze of government policies over the years and the political wrangling behind energy decisions and practices. This is for the good social studies student who is interested in current energy problems and in the types of decisions which must be made to assure our energy future.

181. Hunter, Robert E. **The Energy Crisis and U.S. Foreign Policy.** Headline Series. New York: Foreign Policy Assn., 1973. 80pp. $1.40. Gr. 11–12.

Although parts of this analysis in FPA's Headline Series are dated, this well-written, economically oriented booklet contains worthwhile insights into energy use patterns in the U.S., potential sources of energy imports, and the role of the Soviet Union in the world energy picture. The author concludes that lasting solutions to the energy problem will be found not in technology but in the realm of politics, including the domestic politics of conservation. For sophisticated readers in social studies classes.

182. Linden, Henry R.; Parent, Joseph D.; and Seay, G. Glenn. **Perspectives on U.S. and World Energy Problems.** Chicago, Ill.: Institute of Gas Technology, 1977. unp. free. Gr. 11–12.

This economically oriented manuscript—published by the research and development arm of the gas utility industry and reflecting its viewpoint—is useful for its presentation in capsule form of U.S. and world energy information. Numbers in the table of contents are keyed to separate text blocks dealing with such topics as national differences in the ratio of energy use to Gross Domestic Product, parallel trends in U.S. personal income and energy use, and projections of oil, gas, and uranium resources. Each text block is accompanied by a chart, graph, or table. Selected items from this manuscript will be useful in economics or advanced social studies classes studying contemporary issues; some knowledge of economic terms is necessary.

183. Mancke, Richard B. **The Failure of U.S. Energy Policy.** New York: Columbia Univ. Pr., 1974. 189pp. $11; $3.50 pap. Gr. 11–12.

Mancke asserts that U.S. energy policy has historically been a hit-or-miss effort because of its complex formulation and because decisionmakers have not been able to view the situation in a comprehensive framework. He argues for a national policy which will achieve three goals: preventing environmental pollution, preventing growing dependence on energy from the OPEC nations, and abolishing the percentage depletion granted to the U.S. oil industry. He discusses policies regarding oil imports, natural gas, merchant tankers, superports, emission standards, coal, and nuclear power and recommends a package of energy policy reforms. Scholarly but readable, this could be used by advanced students in government and economics classes.

NONPRINT

184. The Arab World: Oil, Power, Dissension. 1 filmstrip w/1 cassette, 68 fr., color, w/tchr's. guide. Current Affairs, 1974. $22. Gr. 7–12.

Entire nations like Japan, and entire regions like Western Europe, depend on Arab oil for over 70% of their petroleum needs. Oil-price rises of 470% contribute to worldwide inflation. This strip provides background on the Arab nations and their influence, concluding that because of oil the Arabs have a key to vast economic power which poses a change to world order. Purchasers should be aware that several statements in the set are no longer correct because of political changes since 1974 and that any program on the Middle East will probably require updating by the teacher or student using it.

185. The Energy Crisis: Depleting the World's Resources. 1 filmstrip w/1 cassette or 1 disc, 64 fr., color, w/tchr's. guide. Current Affairs, 1973. $22. Gr. 7-12.

The title of this strip is somewhat misleading since it is not as concerned with worldwide energy depletion as it is with the United States's growing dependence on other nations for energy—and the political ramifications of this dependence. It recognizes America's need to maintain its current rate of growth while pointing to the political and economic hazards of energy dependence and the environmental objections to the production of coal, natural gas, and oil. Cursory reference is made to alternate energy sources, but oil and gas are presented as those which will solve our immediate crisis. The judicious use of traditional sources and new discoveries is suggested as a solution to this crisis. The viewer is left with the question: if a choice has to be made between environmental concerns and ensuring that America meets her energy needs, which would you prefer? Primarily for social studies classes.

186. Commander or Captive? Humans in High Energy Society. 1 filmstrip w/1 cassette or 1 disc, 55 fr., 15 min., color, w/tchr's. guide. Multi-Media Productions, 1974. $11.95. Gr. 9-12.

Taking a socioeconomic approach to the energy problem, this filmstrip suggests that though the source of our technology is human energy working on the materials of the earth, a small coterie makes decisions on energy, whether for reasons of profit or to satisfy the demands of consumers who may or may not need high-energy items. It exhorts the viewer to consider the attitudes of low-energy consumers such as the American Indian.

187. Earth, Engines and Electricity. 1 filmstrip w/1 cassette or 1 disc, 57 fr., 15 min., color, w/tchr's. guide. Multi-Media Productions, 1974. $11.95. Gr. 9-12.

Contrasting it with American Indian society, this sound filmstrip gives the background of our high-energy society from its beginning in the English coal mines to the development of the steam engine and the discovery of oil. It poses questions about the validity of our energy-intensive economy. Useful with senior high social studies classes.

188. Global Emergency: Energy (Set). 2 filmstrips w/2 cassettes, 89–91 fr., 19 min. ea., color, w/tchr's. guide. Guidance Associates, 1975. $52.50 set. Gr. 10–12.

New Sources New Values defines the crisis and suggests various methods of creating and harnessing energy. Ralph Nader speaks about concrete dangers from nuclear power plants; an economist presents the problems confronting both industrial and semi-industrial nations if there are cuts in the energy supply or if energy prices rise drastically; a computer simulation scientist speculates that even a new, abundant supply of energy might change our climate and pollute our atmosphere and predicts that finite energy supplies must result in reduced consumption and an altered way of life. *The Energy-Aware Society* provides insights into the energy-wasteful car, symbolizing a society of planned obsolescence, while an expert from the Princeton Center for Alternative Futures suggests a transformation of the economy from energy-intensive to labor-intensive. Two architects discuss the possibilities of alternate housing and an alternate-design city. The final segment looks at the pros and cons of constructing an oil refinery in economically depressed Eastport, Maine. Although somewhat disjointed, the set is nonetheless useful in social studies classes focusing on 20th Century issues and can help students to understand the relationship of energy consumption to economic, environmental, population, and political questions. Pauses in each strip allow class discussion of issues raised.

189. West Virginia: Life, Liberty and the Pursuit of Coal. 16mm, 53 min., color, w/tchr's. guide. Xerox Films, 1973. $595. Gr. 10–12.

Based on the ABC News Closeup from 1973, this long documentary exposes the coal-mining situation in West Virginia, a state which produces 122 million tons of coal a year but still ranks 45th in per capita income. The viewer is acquainted with strip mining and the resulting spoil banks and erosion which leave many "hollows" in danger after even an ordinary rain. The problems of unsafe, unlicensed coal-waste dams created as a result of refining coal are shown to threaten life and property in many areas of the state; official promises to do away with these dams are said to have been unfulfilled. The people are described as apathetic about the dangers or improvement of conditions because they believe there is a close alliance between the state's government and the coal industry. Dozens of interviews with state and federal officials are included, leaving the viewer with the impression of vast hedging, buck-passing, and, possibly, simple corruption. ABC correspondents take up the question of enforcing the 1969 Mine Health and Safety Acts and find that as of 1973 there were 91,000 unpaid fines and almost $20 million outstanding as a result of violations of this act. The conclusion: the technology is available to correct these problems, but the laws are not being enforced and not being complied with by many coal companies. This film will be most useful in social studies classes studying contemporary issues and the functions of state and federal government.

Engines

NONPRINT

190. Introduction to Steam Engines. 1 silent filmstrip, 29 fr., color. Standard Projector and Equipment Co., n.d. $7. Gr. 6–10.

The strip successfully develops the following themes: the history of steam engines, how a steam engine works, the two major types of steam engines, and early and modern uses of steam engines. A lead frame introduces new terms such as piston, flywheel, and turbine. The strip provides diagrams of the piston steam engine and the turbine and offers a thorough explanation of how each has been used in the past and how they are currently used. Diesel and internal combustion engines are mentioned but not explained in detail. Useful as background for more complex filmstrips on power generation.

191. Thermal Engines (Series). 3 ¾" videocassettes or 3 ½" or 1" videotapes, 30 min. ea., color. Nebraska Educational Television Council for Higher Education, 1976. $225 ea.; $35 ea./7 days rental. Gr. 11–12.

This three-part educational-TV set relates the design, use, and environmental effects of engines to their energy use and the energy problems facing this country. In Part I, *How They Work*, Dr. Spencer Sorenson of the University of Illinois at Urbana-Champaign explains with clear diagram sequences how thermal engines, the backbone of our transport system, work. Five types of internal-combustion engines are shown: the four-stroke spark ignition engine, the four-stroke diesel engine, the two-stroke spark ignition engine, the rotary (Wankel) engine, and the gas turbine. The external combustion engine, the Rankine cycle steam engine, and the Stirling cycle engine are also discussed. In Part II, *Applications*, Dr. Sorenson discusses eight factors which determine the choice of engine: cost, fuel consumption, engine weight, size, multi-fuel use, noise, cooling systems, and engine emissions. Using a studio set, still photographs, and some film footage, this presentation is not visually exciting but does contain concrete information about engines and energy consumption. Of particular interest are the reasons why different engines are used in cars, trucks, buses, and airplanes and Sorenson's observation that cars may continue to have gasoline engines despite rising fuel costs because the engines themselves are less costly than alternative engines. In Part III, Sorenson notes that motor vehicles contribute about 70% of the carbon monoxide emissions, 50% of the nitrogen and hydrocarbons. He considers attempts to lower emissions by leaner carburetion, lower compression ratios, retarding spark timing, catalytic converters, etc. He draws comparisons before government controls and after, noting what effects modifications have had on auto performance and fuel usage. This series could be useful in auto mechanics and engine repair and maintenance classes and in some physics units.

Gas

PRINT

192. American Gas Association. **Natural Gas Serves Our Community.** Arlington, Va.: American Gas Assn., n.d. 17pp. inquire about price. Gr. 2-4.

This two-color booklet provides a basic introduction to natural gas: what it is, where it comes from, how it is used. A more detailed version is available for grades 5-9 (see No. 197).

193. American Gas Association. **The History of Natural Gas.** Arlington, Va.: American Gas Assn., n.d. 16pp. inquire about price. Gr. 3-6.

In comic book format, this title tells the story of the gas industry.

194. American Gas Association. **What Happens When You Turn on the Gas.** Arlington, Va.: American Gas Assn., n.d. 28pp. inquire about price. Gr. 4-8.

This is a photo-essay which tells the story of natural gas from the well to the user.

195. American Gas Association. **What Is a Gas?** Arlington, Va.: American Gas Assn., n.d. unp. inquire about price. Gr. 4-8.

This collection of 12 classroom demonstrations/experiments explains, for example, what a gas is like, incomplete burning, and how to make carbon dioxide. It could be used in grades 4-6 or with slow junior high groups.

196. Ridpath, Ian, ed. **Man and Materials: Gas.** Man and Materials Series. Reading, Mass.: Addison-Wesley, 1975. 33pp. $4.95 plb. Gr. 5-8.

Fact-filled and with the same structured format as others in this series, this heavily illustrated book discusses both natural gas and gas manufactured from coal, as well as natural gas substitutes. Exploration, transportation, storage, use, and distribution of natural gas are shown as is the ideal "total energy system" in which natural gas supplies all the power needed in a building, minimizing waste heat.

197. American Gas Association. **Natural Gas Serves Our Communitiy.** Arlington, Va.: American Gas Assn., n.d. inquire about price. Gr. 5-9.

The content of this four-color booklet belies the title, for it is a good reference source which explains where natural gas comes from, how it is found, the drilling process, the pipelines and pumping which transport it, underground storage, regulator stations, and uses in the community. A self-test is included.

198. American Gas Association. **The Story of Natural Gas Energy.** Arlington, Va.: American Gas Assn., n.d. 48pp. inquire about price. Gr. 6–12.

This is a detailed explanation of the resource, where it comes from, how it is routed to customers, and its uses. A short chapter discusses reserves and exploration.

199. American Gas Association. **Science Principles and Gas Appliances with Experiments.** Arlington, Va.: American Gas Assn., n.d. 32pp. inquire about price. Gr. 9–12.

This booklet discusses the science principles used in six gas appliances: range, home heater, water heater, clothes dryer, incinerator, and air conditioner. Twelve classroom experiments and materials for overhead projection are included. For senior high science classes.

200. American Gas Association. **Experiments: Properties of Gas and Heat Energy.** Arlington, Va.: American Gas Assn., n.d. unp. inquire about price. Gr. 10–12.

This publication contains 17 experiments on properties of gas (e.g. Boyle's Law, the speed of sound in a gas) and 14 on heat energy (steam to water, how fast gas heats water). These are prepared on spirit masters and could be used in physical chemistry or slow chemistry classes in the senior high schools.

NONPRINT

201. Natural Gas—Science behind Your Burner (Kit). 1 filmstrip, 37 fr., 2 charts, tchr's. guide. American Gas Assn., n.d. inquire about price. Gr. 4–7.

A kit which consists of one 37-frame filmstrip; two wall charts, one showing natural gas pipelines in the U.S. and the other the flow of gas from well to burner; and two spirit masters. The filmstrip traces the gas backward from a stove to the meter, the gas main, the local regulatory station, and a pipeline to the gas well. The kit includes a teacher's guide and is useful for science classes where reading skills are being encouraged.

202. Natural Gas Serves Our Community (Kit). 28 pictures. American Gas Assn., n.d. inquire about price. Gr. 4–7.

This cutout kit consists of 28 pictures on pasteboard in four colors and ribbon to represent pipelines. It depicts the story of gas from the fields to the community. The kit includes the middle-school version of the text of the same name (see No. 197).

203. Geology and Natural Gas. 16mm, 15 min., color, w/tchr's. guide. Cavalcade Productions, 1968. $130; $35/week rental; free loan for those in Northern Ill. Gas Co. service area. Gr. 4–8.

Using a dialogue between himself and his dummy Speedy as the framework, ventriloquist Bob explains where natural gas comes from and how we get it. Lab demonstrations using a silent teenager, Joan, show that gas is found in porous sandstone or limestone and how a seismograph is used to determine the location of gas fields. Field footage shows how deep wells are sunk thousands of feet underground, sample borings are brought up and checked microscopically, pipeline is laid to transport gas from the Southwestern states to consumers, and surplus gas is stored in aquifers for use in the peak-demand season. Dialogue at the end includes the main points to be learned from the film.

204. More to Come. 1 filmstrip w/1 cassette, 46 fr., 5 min., w/tchr's. guide. Educational Services, American Gas Assn., 1977. inquire about price. Gr. 4-8.

A nontechnical introduction to the idea of "proved reserves" of natural gas and the fact that use of gas has outstripped reserves. This film briefly shows what the natural gas industry is doing to find new reserves—bringing liquefied gas from other parts of the world and experimenting to produce gas from coal. It shows how we use gas and asks, "Are we using energy wisely?" The basis for an intelligent answer is not provided within the context of the strip.

205. The Physical and Chemical Properties of Natural Gas. 16mm,11 min., color, w/tchr's. guide. Cavalcade Productions, 1968. $150; $40/week rental; free loan for those in Northern Ill. Gas Co. service area. Gr. 4-8.

A clear, direct dialogue between ventriloquist Bob and his dummy Speedy, this film uses simple laboratory demonstrations to show half-a-dozen properties of natural gas: it is colorless; odorless until odor is added; nontoxic; lighter than air; needs oxygen for combustion; and is composed principally of the two hydrocarbons, methane and ethane. The silent demonstrator is a teenager named Joan who puts a guinea pig in a container of natural gas and air; fills a balloon; and mixes oxygen and natural gas in a welding torch, among other experiments.

206. How Your Gas Meter Works (Kit). 17" x 22" wall chart, spirit masters, 4-page brochure. American Gas Assn., n.d. inquire about price. Gr. 9-12.

This kit includes a very detailed 17" x 22" wall chart, a four-page brochure explaining the workings of a gas meter, and spirit masters called "Reading Your Meter" and "How Your Gas Meter Works."

207. Nuclear Gas Stimulation—Tapping Our Natural Heritage. Energy Sources: A New Beginning Series. 3/4" videocassette or 16mm, 29 min., color, w/tchr's. guide. Univ. of Colorado, Educational Media Center, 1975. $230 videocassette; $333 16mm; $2.50 tchr's. guide; $10/3 days, $20/15 days, $30/30 days videocassette rental. Gr. 10-12.

Natural gas provides 31% of our energy. Excluding transportation purposes, it is our most popular fuel because it is less polluting and cheaper. But by the year 2000, demand is expected to greatly exceed supply, so areas thought to have gas in rocks with low permeability must be tapped. The narrator first uses a laboratory experiment to illustrate what fracturing these "tight" rocks makes possible, then explains the three methods by which gas in these rocks can be released: drilling, hydraulic fracturing, and explosive fracturing by conventional or nuclear methods. Much of the film is devoted to the success and problems of nuclear fracturing, with the process shown in animated diagrams. At the time it was produced, there were no real answers to the questions raised by this film, i.e., is nuclear fracturing technologically possible? commercially feasible? because of radiation from the explosions, is it safe? Somewhat static studio shots are relieved by on-site footage of the government's experimental Rio Blanco detonations. Detailed information is presented in educational-TV style.

Geothermal Power

PRINT

208. Ravenholt, Albert. **Energy from Heat in the Earth.** Hanover, N. H.: American Universities Field Staff, 1977. 10pp. $1.50; $.75 to subscribers. Gr. 10-12.

One of the Fieldstaff Reports on major contemporary issues, this work is subtitled "Philippines Taps Immense Resources of Pacific Fire Belt." Ravenholt describes the background and activities of a project that by 1985 should provide 24% of the Philippines' electric generating capacity and save that country 180 million U.S. dollars annually in fuel oil imports.

209. Ravenholt, Albert. **Wairakei and New Zealand's Thermal Power.** Hanover, N. H.: American Universities Field Staff, 1977. 9pp. $1.50; $.75 to subscribers. Gr. 11-12.

One of a series of Fieldstaff Reports on international and major contemporary global issues, this states that the success of New Zealand's generation of electric power using geothermal wet steam not only is promising for New Zealand's future energy needs but has significant implications for 24 countries whose geography would also be conducive to tapping the Pacific Fire Belt. It describes briefly the history, effectiveness, and cost of geothermal use in New Zealand.

210. Geothermal Power: The Great Furnace. Energy Sources: A New Beginning Series. 3/4" videocassette or 16mm, 29 min., color, w/tchr's. guide. Univ. of Colorado, Educational Media Center, 1975. $230 videocassette; $333 16mm; $2.50 tchr's. guide; $10/2 days 16mm rental; $10/3 days, $20/15 days, $30/30 days videocassette rental. Gr. 10-12.

This film is probably the most technical and quickly paced in the *Energy Sources:A New Beginning* series. It asks, is geothermal energy (the heat stored below the earth in rocks or water and steam-filled pores of rock masses) available, efficient, and economically feasible? Using the same lecture format as others in this series, the narrator explains the relationship between continental plates and likely geothermal areas such as the western coasts of North and South America. There is outdoor footage featuring Dr. David Kassory of the University of Colorado. Against the background of a geothermal swimming pool and the snow-capped Rockies, he explains the techniques for determining the extent, quality, and duration of a geothermal reservoir. The drilling systems for utilizing steam and hot water-dominated geothermal energy are clearly charted, and the potential for hydro-fracturing dry geothermal energy or using a closed binary system which returns steam underground are discussed. The one operating geothermal generating plant in the U. S., supplying a third of San Francisco's power, is shown and described. Problems associated with geothermal—the sulfurous odor and the need to place the generating plant very close to the heat source—are examined. The film notes that several federal agencies disagree radically as to the potential contribution of geothermal to the energy picture.

211. Geothermal Energy. ¾" videocassette or 1" or ½" videotape, 29:25 min., color. Nebraska Educational Television Council for Higher Education, 1973. $225; $35/7 days rental. Gr. 11-12.

A somewhat technical but interesting television lecture on what geothermal energy is and where it is found throughout the world. It suggests that geothermal energy may be more widely distributed than suspected. Types of thermal areas are discussed: the normal gradient from which low grade heat for space heating is produced, and the hypothermal type field whose energy is used for power generation. There is interesting footage of a steam field in Central America where wells have been actively developed since the 1960s. The program shows the complex exploration for geothermal energy and the equipment used for actual drilling and experimental production. Sites in New Zealand and Iceland are shown, and the lecturer notes that geothermal resources may be more intensely developed as prices of competitive fuels rise. The environmental impact, including corrosive hydrogen sulfide gas in the steam and the upset of the ecological balance if spent geothermal steam is dumped into nearby bodies of water, is acknowledged. This low-key production will be most useful with advanced science classes studying energy alternatives.

Heat Pump

NONPRINT

212. Bill Loosely's Heat Pump. 16mm, 10 min., color. National Film Board of Canada, 1976. Dist.: Bullfrog Films. $140; $15/3 days rental. Gr. 9-12.

A National Film Board of Canada film, this production features a neat, quiet tour of a heat-pump system of home heating designed by an Ontario engineer when he built his house more than 25 years ago. The system takes heat out of the ground surrounding his house and uses it to heat freon, which is vaporized and fed into a heat-exchanger which yields hot air to heat the house. Loosely's original cost was $1200; today he estimates the cost would be $4000. Animated diagrams of how the system works are particularly good. Most useful in physics classes or with general science students who have some understanding of physics, it might also be used in science classes studying alternate energy sources.

Methane

NONPRINT

213. Bate's Car. 16mm, 15:33 min., color. National Film Board of Canada, 1974. Dist.: Arthur Mokin Productions. $230; $30/2 days rental. Gr. 10-12.

An ingenuous but sane encounter with Harold Bate who has found his own solution to the gasoline shortage in a barnyard manure pile. Bate, a classic eccentric inventor, mixes pig and horse manure, straw, and water to produce methane gas. He has adapted his car to run on either gasoline or methane and extols the virtues of methane: clean, no-knock, no exhaust except slight water vapor, and 98% complete combustion, far higher than that of gasoline. Equipped with a converter under the hood, the car can switch from gasoline to methane with ease. Bate, who has received more than 10,000 letters from all over the world, is also working on an electric car and a bicycle that will go up to 40mph and is partly self-propelled. Meanwhile, he praises methane as a reliable, cheap source of energy. This film could be used in current events discussions in social studies and economics classes to illustrate an offbeat solution to the energy crisis, in senior high chemistry classes studying methane, and in physics classes studying energy. It could also be used with bright junior high students who can cope with Bate's rather muttered English.

Methanol

PRINT

214. Lincoln, John Ware. **Methanol and Other Ways around the Gas Pump.** Charlotte, Vt.: Garden Way, 1976. 134pp. $8.95; $4.95 pap. Gr. 9–12.

Lincoln makes a good case for the use of methanol (wood alcohol), or methanol blended with gasoline, as a practical fuel for automobiles. He describes research and tests with methanol and explains why there has been resistance to a methanol industry. Several chapters take up the history of alternate fuels and engines. A lively, well-researched popular science work, this could be used by chemistry or general science students or by any student curious about energy alternatives.

Nuclear Energy

PRINT

215. Michelsohn, David Reuben. **Atomic Energy for Human Needs.** New York: Messner, 1976. 189pp. $7.29 plb. Gr. 5–8.

Focusing on peacetime uses of atomic energy, this work covers radiocarbon dating and the use of the atom in agriculture, mining, medicine, construction, and industry. Chapter five takes up the question of using atomic power to generate electricity, implying that it is cheap, nonpolluting (at least in the traditional sense), and easy to transport. Michelsohn does, however, explore the dangers involved in nuclear accidents and in waste disposal. He provides a good basic description of the fusion process and the problems surrounding its development for commercial purposes.

216. Atomic Energy. Merit Badge Pamphlet No. 3275. North Brunswick, N. J.: Boy Scouts of America, 1965. 68pp. $.55 pap. Gr. 6–9.

The purpose of this pamphlet is to prepare Scouts and Explorers to meet the requirements for the Atomic Energy merit badge. It provides a brief overview of atomic history; an explanation of atomic structure; clear explanations of critical mass, nuclear fission and fusion; labelled photographs of a model reactor and a list of materials needed to make a reactor (one of the optional requirements for the badge); a section on peaceful uses of atomic energy ; sections on radiation and on instruments for detecting radiation; and a final section on careers in the atomic energy industry. The detailed glossary is particularly useful because of the technical subject matter. This pamphlet is comprehensive and well presented, but it is now 13 years old; it should be updated to provide current addresses and references and to present a more balanced picture of how atomic energy can be used for electrical power generation.

217. Adler, Irving. **Atomic Energy.** The Reason Why Series. New York: John Day, 1971. 48pp. $5.50 plb. Gr. 7-9.

The format and drawing style of this work first suggests that it is an upper elementary title on how we get energy from the atom and what is done with it. Actually the concepts, e. g. atomic structure and fissioning, generally are better understood at the junior high level. Brief, useful information is included on topics such as reactor design and preparing nuclear fuels. The text is not divided so the table of contents must be consulted.

218. Asimov, Isaac. **How Did We Find Out about Nuclear Power?** How Did We Find Out Series. New York: Walker, 1976. 64pp. $5.85 plb. Gr. 7-9.

Asimov recounts the historical discoveries which unlocked the secrets of atomic energy and led to its use as a power source, and he describes the achievements of each scientist involved in the effort. In the final chapter he explains atomic fission and fusion and notes the dangers of breeder reactors and the problems surrounding controlled nuclear fusion. Terms are explained within the text, but there is a great deal to absorb; use will probably be limited to the more advanced science students in grades 7-9.

219. Atomic Industrial Forum. **Nuclear Brochure Series.** Washington, D.C.: Atomic Industrial Forum, 1976 & 1977. 8pp. single copies free; bulk requests, $.08 per brochure. Gr. 7-12.

The trade association for the nuclear power industry offers ten illustrated brochures presenting their subjective position and statistics on the following nuclear energy topics: How Nuclear Plants Work, Nuclear Reactor Safety, Managing Nuclear Waste, The Savings with Nuclear Energy, Shipping Nuclear Fuel, Plutonium in Perspective, Recycling Nuclear Fuel, Protecting Nuclear Power Plants, Uranium: Energy for the Future, Insuring Nuclear Risks.

220. Benrey, Ronald M. **Nuclear Experiments You Can Do.** Southfield, Mich.: Thomas Alva Edison Foundation, 1976. 32pp. $25/100; $21/100 for orders of 500 copies or more. Gr. 7-12.

The eight experiments in this booklet vary considerably in terms of complexity. Some require using both alpha and gamma rays, and the author suggests a low-cost source for them. Experiments include observing radioactivity through radiography and a cloud chamber, demonstrating how radioactivity can be shielded, and building a geiger counter. For junior and senior high science classes.

221. Olson, McKinley C. **Unacceptable Risk: The Nuclear Power Controversy.** New York: Bantam, 1976. 309pp. $2.25 pap. Gr. 9-12.

The title is a take-off on Dr. Ralph Lapp's statement that in the face of dwindling fossil fuels and the relatively undeveloped state of alternate energy sources, nuclear power is "an acceptable risk." Olson describes the development of the nuclear industry and the growth of the anti-nuclear movement among scientists, environmentalists, and workers in the industry. He discusses the dangers of plutonium, the problems in the nuclear fuel cycle, the reasons why nuclear plants are more expensive and less efficient than promised, the industry's attitudes, and the massive slowdown in nuclear development since 1972. Though advertised as "the only book that tells all sides of the nuclear energy story," its sympathies and coverage are clearly with the opposition. An appendix provides a list of operating and proposed nuclear power plants in the U. S.

222. Stencel, Sandra. "Nuclear Waste Disposal." **Editorial Research Report.** Vol. 2, No. 21. Washington, D. C. : Editorial Research Reports, 1976. 22pp. $2.50. Gr. 9-12.

Citing a variety of sources, the author takes up the controversial topic of nuclear waste disposal in three sections: management of radioactive wastes, search for waste disposal solutions, and the future of plutonium reprocessing. She describes efforts to find a permanent method of storing wastes up to the fall of 1976. A bibliography of articles, studies, and reports is appended.

223. Committee for Economic Development. Research and Policy Committee. **Nuclear Energy and National Security: A Statement on National Policy.** New York: Committee for Economic Development, 1976. 78pp. $4 plb.; $2.50 pap. Gr. 10-12.

The Committee for Economic Development is an independent research and educational organization of 200 business executives and educators. In this statement, the Council asserts that it is not within the power of the United States to turn the world away from nuclear energy and that only by continuing to exercise leadership in the worldwide development of atomic energy can the U.S. safeguard the world's nuclear economy. The Council makes specific recommendations as to how the U.S. can take this lead. In addition, it comes out strongly for a national energy policy which recognizes that domestic and foreign energy policy are inseparable and that energy policy is part of national security policy for which "the highest level of presidential authority," not Congress, should be responsible. A chart of world nuclear power plants in operation or under construction as of December 31, 1975, is provided in the appendix.

224. Hayes, Denis. **Nuclear Power: The Fifth Horseman.** Worldwatch Paper No. 6. Washington, D. C.: Worldwatch Institute, 1976. 68pp. $2 pap. Gr. 10-12.

This paper evaluates the future of nuclear power, subjecting it to tests of economy, safety, adequacy of fuel supplies, environmental impact, and national and international security. Hayes puts nuclear energy in worldwide perspective, noting that interest in nuclear energy has ebbed in the Western world but that fierce competition for the nuclear market has evolved in the developing nations, and that no country has yet devised an adequate solution to the problem of eliminating radioactive wastes. He concludes in the section on safety that: "it is impossible to eliminate all risk. The level of acceptable risk is an ethical rather than a technical matter. Consequently, the final decisions are not scientific, but are, rather, social and political." A four-page section on "Nuclear Power and Society" examines nuclear power in relation to the total energy scheme. Somewhat technical by nature, this work provides a good summary of the issues surrounding nuclear power for the social studies student with some scientific background. This book is adapted from Hayes's *Rays of Hope: The Transition to a Post-Petroleum World* (see No. 175).

225. McPhee, John. **The Curve of Binding Energy.** New York: Farrar, 1974. 170pp. $8.95. Ballantine, $1.95 pap. Gr. 10–12.

The title refers to the curve which tells how much energy may be extracted from the fission of a single atom. The author describes in a rambling fashion a tour of an American nuclear facility conducted by Dr. Theodore Taylor, a physicist who worked on atomic bomb design at Los Alamos. Taylor feels that fissionable materials could easily be stolen from nuclear installations. The reader is left with the feeling that the risks involved in any fission program are so great that the programs should be abandoned in favor of increased fusion research projects. Parts of the book were published as a series in the *New Yorker* in 1973. Students who are science buffs will be the likely audience.

226. Nader, Ralph, and Abbotts, John. **The Menace of Atomic Energy.** New York: Norton, 1977. 414pp. $10.95. Gr. 10-12.

As the title indicates, this thoroughly documented but generally nontechnical work contends that nuclear power is unacceptable as a source of energy for present or future societies. It focuses on specific issues and technical problems, among them radiation effects, reactor safety, environmental effects of reactor operation, reprocessing and nuclear wastes, worker safety, and the plutonium breeder reactor. The authors claim that nuclear power is now an economic disaster—a "technological Vietnam." They examine the atomic industry, the federal push behind it, and the international spread of atomic power. The closing section suggests specific methods of citizen action with Congressmen, the Nuclear Regulatory Commission, and the public utilities at the state and community level. The appendix contains a list of citizen groups and governmental sources which provide information on nuclear energy issues and strategies. Best used with social studies and contemporary issues students who share Nader's tenacity.

227. Novick, Sheldon. **The Electric War: The Fight Over Nuclear Power.** San Francisco, Calif.: Sierra Club Books, 1976. 376pp. $12.50. Gr. 10–12.

The author states that this is "an attempt to display as much of the battle as can be comfortably viewed, enough for a reasonable person to form an opinion about nuclear power." Through interviews, Novick allows the protagonists—uranium miners, physicists, utility executives, members of regulatory agencies, environmentalists—to speak for themselves. Contending that three great forces—the Cold War, the struggle between governments and corporations, and the creation of the private electric power companies—have shaped the nuclear power industry, he sets these interviews against historical vignettes. This is a fascinating but fragmented book which must be read from cover to cover to get a complete picture. The best audience will be curious social studies students.

228. Scientists' Institute for Public Information. **Nuclear Power, Economics and the Environment.** New York: Scientists' Institute for Public Information, 1976. 73pp. $2. Gr. 10–12.

One of five Science Topic Kits on energy-related subjects, this work brings together seven articles, from 1974 or later issues of *Environment* magazine, on subjects such as the possibility of a reactor explosion, hot wastes from nuclear plants, and reprocessing nuclear fuels. The introduction provides background on the federal government and private industry investment in nuclear power and suggests that the public should become involved in energy policy making. Other Science Topic Kits on energy, priced at $2.50 to $3.50, include *Alternative Sources, Consumption and Conservation, Fossil Fuels,* and *Occupational Hazards.* For secondary students with some chemistry background.

229. Webb, Richard E. **The Accident Hazards of Nuclear Power Plants.** Amherst, Mass.: Univ. of Massachusetts Pr., 1975. 228pp. $6.95 pap. Gr. 10–12.

Although the main part of this book is too technical for most high school students, the "Summary Description of Fourteen Accidents and Near Accidents in Nuclear Reactors" included in the appendices is the only article seen which provides the facts about nuclear accidents. The effect is impressive. The acronyms in the appendix are clarified in the introduction and in the chapter references and notes.

230. Willrich, Mason, and Lester, Richard K. **Radioactive Waste Management and Regulation.** New York: The Free Press, 1977. 130pp. $13.95. Gr. 10–12.

Based on a Massachusetts Institute of Technology research project, this work examines the character of radioactive waste from both commercial and military sources and discusses the range of decisions involved in waste management. The authors recommend specific institutional reforms which they feel are necessary for the safe management of radioactive wastes both in the U. S. and abroad. Though the focus of this study is not overly technical, it is specialized and will be most useful as a resource for special assignments by more advanced science students.

231. Union of Concerned Scientists. **The Nuclear Fuel Cycle: A Survey of the Public Health, Environmental and National Security Effects of Nuclear Power.** MIT Press Environmental Studies Series. rev. ed. Cambridge, Mass.: MIT Pr., 1975. 291pp. $4.95 pap. Gr. 11-12.

An official U.S. Atomic Energy Commission assessment of the hazards associated with the nuclear fuel cycle appeared in 1972. The Union of Concerned Scientists felt that the AEC document was marked by serious errors and omissions and initiated this independent, parallel study. Separate chapters review the exposure of uranium miners to radioactive, carcinogenic gas; the milling of uranium and the disposal of tailings; the mechanisms and consequences of catastrophic accidents; the problems of diversion of nuclear materials for terrorist purposes; the hazards of transport; the history of operations of the first American fuel reprocessing plant; and the storage/disposal of high-level radioactive wastes. An overview chapter calls for full public understanding of the special features of nuclear energy and a moratorium on nuclear plants until the problems surrounding it have been resolved. Although many of these articles suffer from the verbiage of science reports, they are not beyond the reach of very bright secondary students. A detailed index makes the contents accessible for reference purposes.

NONPRINT

232. Introducing Atoms and Nuclear Energy. 16mm, 11 min., color. Coronet Films, 1963. $160. Gr. 5-9.

Atomic structure is explained and illustrated as a background to understanding the process by which nuclear energy is released. The chain reaction which takes place in nuclear fission is demonstrated, and nuclear fusion in the sun is explained in animated sequences.

233. Clinch River Breeder Plant Project. 100 slides w/script, 45 min. Oak Ridge Associated Universities, Energy Education Division, 1976. long-term free loan. Gr. 6-12.

This slide set was designed to present the position of the Nuclear Reactor Corporation (the U. S. electric systems supporting the Clinch River Breeder Reactor Plant Project) on the potential and safety of the breeder reactor for electrical power generation. Slides are furnished in a loose-leaf notebook. The complete set can be shown or sections from it may be combined under the following topics: General Energy Situation, Nuclear Power, The Breeder

Reactor, Clinch River Breeder Reactor Plant, Plutonium, and Reactor Safety. Many of the slides contain diagrams or printed information. A 27-page script provides the accompanying narrative. Availability of this set will depend on the continuance of the project, since it is also funded by the federal government.

234. Energy and Nuclear Power. 16mm, 14 min., color, w/tchr's. guide. Xerox Films, 1977. $240. Gr. 6-12.

First noting that America is an energy-oriented, energy-wasteful nation, this film examines one solution to the energy shortage: the use of nuclear power for generating electricity. Nuclear's advantages—the small amount of fuel, savings on transporting large amounts of fuel to the generating plants, lack of obvious pollution—are set forth, as are the safety back-up systems for these plants. The last half of the film, however, details the problems of nuclear: what to do with long-lived nuclear wastes, a "monumental legacy"; the health dangers of uranium mining; the problem of what to do when the uranium runs out and nuclear plants must be shut down. It allows the viewer to decide if the advantage of nuclear reactors and their inherent level of risk make them an acceptable solution to the energy problem. For science and social studies classes at both junior and senior high levels.

235. Atomic War/Atomic Peace: Life in the Nuclear Age (Series). 6 audio-cassettes, 9-25 min., w/tchr's. guide. Visual Education Corp., 1975. $67 ser. Gr. 7-12.

This program documents man's discovery of the principles behind fission and fusion, the development of the A-bomb, the bombing of Japan, nuclear testing and the Cold War, the peaceful atom, and the debate over the safety of nuclear power plants. Historical figures like Einstein and Eisenhower talk about research on the atom and its potential for war and peace. Of particular interest is cassette five, which recalls the launching of the first nuclear submarine; the First International Conference on Atomic Energy in 1955, with its prediction of growing world energy needs; and the opening of the first atomic power station at Shippingport. In cassette six, the issue of the safety of nuclear power plants is debated in interviews with Dr. Saul Levine of the Energy Research and Development Administration and David Brower of Friends of the Earth. Levine claims that massive research has gone into the storage and transport of radioactive waste products from spent nuclear fuel and that their hazard has been overstated. Brower challenges America to re-evaluate the decision to develop atomic power and to rethink the priorities that seem to demand an almost frantic growth in energy consumption, further polarizing the have and have-not nations. He calls for laws which would require proof of safety of atomic installations before they are allowed to function. He advises a turn from nuclear to solar energy research and claims that the country's massive investment in nuclear generation of power has held back solar research. Brower considers the energy problem a social rather than a scientific one and believes it should be settled by all Americans.

236. Energy: The Nuclear Alternative. 16mm, 20 min., color, w/tchr's. guide: Churchill Films, 1974. $280; $21/3 days rental. Gr. 7-12.

A nontechnical approach to nuclear energy, this film focuses on the controversy regarding the safety of nuclear generation of electricity. Scientists and concerned citizens clash with the official opinions of the then Atomic Energy Commission and the interests of utilities and private companies building reactors. The film concentrates on three problems: reactor safety and the catastrophic effects of a reactor meltdown, the transportation of nuclear materials, and the disposal of radioactive wastes. Short interviews with nuclear scientists and utility executives, including one with Dr. Edward Teller who says nuclear plants should be located underground, leave it up to the viewer to decide if the plans to build at least 500 nuclear plants by 2000 should go ahead, or if there should be a moratorium and a redirection of research funds to alternative energy sources until the problems surrounding nuclear energy are surmounted. For social studies or science classes dealing with contemporary social-scientific problems.

237. My Nuclear Neighbors. 16mm, 14½ min., color. Walter J. Klein, Ltd., 1971. $275. Gr. 7-12.

Actor Raymond Burr leads this sincere grand tour of nuclear plants in the U.S. Operating from the premise that electric power companies were seeking a power source that would minimize impact on the environment, Burr explains that current nuclear plants are completely safe, clean, and a positive influence on their neighbors in that they create jobs and tax revenues, reduce pollution, and improve conditions in areas adjoining nuclear plants. Statements asserting, for example, that the only thing seen around nuclear plants are "harmless clouds of condensed water vapor" lead the viewer to believe that there is no dispute over the safety of these plants. Nonspecific references are made to tests which monitor their environmental impact. Sponsored by a group of electric companies and produced in cooperation with the Atomic Energy Council, this film requires an introduction and follow-up to provide a fair picture of nuclear's status as an energy source.

238. Nuclear Power: Pro and Con (Set). 16mm, 2 films, 25 min. ea., color, w/tchr's. guide. McGraw-Hill Films, 1977. $725 set or $385 ea.; $40 ea./1 day rental. Gr. 7-12.

A pair of films which will do little to resolve the question "Is nuclear power safe?" but which can stir up considerable discussion about the issue. Based on an ABC News Closeup and narrated by Howard K. Smith, one makes a case for nuclear power, the other against. Each includes a myriad of interviews with experts stating contrasting views. Dr. John Gofman says the Atomic Energy Commission is deluding the public about the dangers of nuclear power, and his counterpart at Massachusetts Institute of Technology, Norman Rasmussen, claims nuclear plants are specifically designed so that failures will not occur and that other forms of energy production have a far worse accident rate. Those against nuclear power cite the danger of human errors, the problems from nuclear waste epitomized in the leakage of World War II wastes at the government's Hanford reservation, and the nonperformance of the Nuclear Regulatory Commission, which they say includes people from the nuclear industry itself. There is frightening footage on the pollution problem from uranium mining, and they prophesy a deadly legacy for generations to come. Those for nuclear power point to the industry's record in generating power during the severe 1976–77 winter, the obsession of the industry with safety precautions, and the fact that other countries are ahead of us in endorsing the breeder reactor. James Schlesinger claims we *must* use both coal and nuclear energy to meet near-future needs. Smith is neutral, urging the country to come to a decision. Since each of these films viewed alone could be extremely convincing, it is recommended that they be shown back-to-back. The overriding problem is that the viewer is not given, for security or other reasons, enough information to make a rational decision. More appropriate for social science classes discussing contemporary problems than science classes since actual scientific content is thin.

239. Physical Science Topics. General Science, Unit 2. 5 transparencies w/1–5 overlays ea. GAF Corp. $3.25–$7.70 ea. Gr. 7–12.

These overhead transparencies all concern nuclear energy. They are, with the classes in which they could be used, as follows: *Atomic Fission* ($4.35, 1 overlay)—chemistry, physics, physical science; *Chain Reaction* ($7.70, 5 overlays)—chemistry, physics, physical science, and with brighter junior high students doing basic chemistry; *Atomic Fusion* ($3.25, 1 overlay)—chemistry, physics, physical science; *Radiation* ($5.50, 3 overlays)—chemistry, physics, physical science; *Atomic Reactor* ($6.60, 4 overlays)—physics, physical science, and junior high general science. Useful in teaching basic concepts, these transparencies may be a more economical purchase than a filmstrip when augmented with the teacher's narration.

240. U.S. Reactor Types. 4 prints, 8½"x11", b&w. Atomic Industrial Forum, 1974. single copies free; $.08/set of 4 for bulk requests. Gr. 7–12.

Black and white graphics showing the four reactor types used in the U.S. and providing brief information about their operation on the back of each schematic diagram. The 8½"x11" format allows for photocopying or other duplication of these drawings.

241. Energy Crisis: The Nuclear Alternative. 16mm, 52 min., color, w/tchr's. guide. Xerox Films, 1973. $575. Gr. 9–12.

Yet another attempt, albeit five years old, to examine the pros and cons of nuclear energy for generating electric power. ABC News correspondent Frank Reynolds interviews several dozen experts on both sides of the issue and examines all the objections to nuclear energy: the possibility of a reactor meltdown, escape of radioactive gases and material, the problem of storing nuclear waste material. Scientists from a conservation-environmental coalition opposed to AEC policies at the time and a spokesman for a Chicago business group speak against allowing plants to operate at full capacity until more is learned about safety systems. Systems for above-ground storage facilities requiring constant human surveillance and below-ground salt-cavern storage are described. Positions are taken on the breeder reactor and the dangers of plutonium compared to uranium. In the last half of the second part of the film, Dr. Richard Balzhiser, then assistant director of the President's Office of Science and Technology, discusses viable energy alternatives. Correspondent Reynolds weighs the questions of "acceptable risk" vs. "haste can produce the most terrible kind of waste." For science classes interested in the ethical issues surrounding the development of nuclear energy.

242. Exploring Magnetic Fusion Energy. wall chart, 2' x 3', color. Grumman Aerospace Corp., Horizons Magazine. single copies free to principals or superintendents when requested on their letterhead; $.50 ea. to others. Gr. 9–12.

This four-color 2'x3' wall chart explains in words and pictures what the U.S. and the Russians have done to make controlled thermonuclear fusion a practical energy source. It concentrates on research being done at the Princeton Plasma Physics Laboratory, Oak Ridge, Lawrence Livermore Laboratory and Lawrence Berkeley Laboratory, and Los Alamos. The producer and Ebasco Services Inc. are working on the Tokamak Fusion Test Reactor scheduled for completion at Princeton in mid-1981. Useful for physics and chemistry students.

243. Now That the Dinosaurs Are Gone. 16mm, 26 min., color. Atomic Industrial Forum, 1974. $250; free loan for 2 weeks from local utility or AIF. Gr. 9–12.

Produced for and released by the trade association of the nuclear power industry, this film purports to explore "the real issues of nuclear power without the emotionalism with which they are usually shrouded" and to answer the questions: why build nuclear plants, how safe are they, and what effect will they have on the environment and the public. Pro-nuclear experts such as Ralph Lapp, Norman Rasmussen of the Massachusetts Institute of Technology, and former Atomic Energy Commission chairman, Dixy Lee Ray, are interviewed to prove: nuclear-based power is the only sensible source now that fossil fuels are going or near gone (conservation is dismissed as a delaying action, not a solution); all of man's actions have some effect on the earth, but nuclear has less than others; nuclear *is* safe—the nuclear industry has 180 reactor years of safe power generation, is strictly policed, and has had no injury or accident to any member of the public. Rasmussen describes the design barriers against radioactive release in case of a melt-down; Lapp states that the amount of radioactivity continually being released into the environment is insignificant; Ray vaguely describes procedures for the disposal of radioactive waste. She stresses the need for facts which will help the public make the decision regarding use of nuclear power, but all the facts are not presented here.

244. Nuclear Power Inside and Out. 20 slides, color, w/script. Atomic Industrial Forum, 1975. $60; 2 week preview. Gr. 9–12.

A clearly and professionally executed set, with accompanying script—but the price is high. Two slides show operable and under-construction nuclear plants in the U.S. as of 1974. Others are captioned diagrams of the fission process; the boiling water, pressurized water, high-temperature gas-cooled, and liquid metal fast breeder reactors; and photographs of actual nuclear power plants employing these reactor types. The concluding slides are artist's concepts of the Clinch River Breeder Reactor Project, whose demonstration phase has a target date of the mid-1980s, and a proposed offshore nuclear plant.

245. Why Nuclear Energy: A Slide Talk with a Moral. 135 slides, color. Atomic Industrial Forum, 1977. $75; 2 week preview. Gr. 9–12.

An attractive, one-sided presentation by the atomic energy industry, this set uses a combination of Aesop and Thurber fables, cartoons, photographs, and statistics to espouse domestic energy independence through nuclear-generated electricity. It notes that oil, gas, and coal have important industrial uses, while uranium's sole commercial application is to generate electricity. The reliability of nuclear energy is stressed over "so-called alternative energy sources." No sources are given for statistics about the high costs of these alternative sources, nor for the claim that nuclear electricity is 50% cheaper than oil and 18% cheaper than coal. Useful in senior high social studies classes to stimulate discussion on the energy choices open to us, this set demands an understanding of the vested interests in the energy industry.

246. Fusion: The Energy Promise. NOVA Series. 16mm, 52 min., color, w/tchr's. guide. WGBH-TV, Boston, 1976. Dist: Time-Life Multimedia. $550; $60/3 days rental. (videocassette avail. from Public Television Library). Gr. 10–12.

This film effectively traces the post-World War II history of fusion research directed towards generation of power. In Part I, Dr. David Rose of M.I.T. explains the problems of producing controlled fusion to a group of English schoolchildren; he uses balloons and the speed of contact between them to indicate the difficulty of the scientific problem. Raw materials such as deuterium and tritium must be transformed into a fourth state of matter called plasma, "a luminous soup of nuclei," a state of matter rare on earth. The sun, which achieves controlled fusion, runs comfortably at a temperature of about 20 million degrees. Because on earth the reactor is small, the temperature will reach 200 to 1,000 million degrees, making the plasma too hot to handle; it must be confined by magnetic fields. Animated cartoons introduce the Great Reactor Race complicated by security clampdowns. In Part II, the overriding technical problem—keeping the plasma within the magnetic field—is explored. Scientists from the U.S., Britain, Russia, and a European consortium tackle the problem. They are inspired by the excitement of working in a totally new field—seeking a scientific holy grail—and the desire of many scientists to put fusion to work for peaceful purposes. The work on the British ZETZ is described, as is the apparent breakthrough by Russia with TOKAMAK, which changed the direction of the U.S. research. The film concludes with recent research using laser fusion to squeeze a pellet of deuterium and tritium in order to produce fusion and emphasizes the need for international cooperation if this fuel source is to be exploited. It also notes some of the possible effects on the environment if controlled fusion is achieved on a wide scale. For upper level science classes.

247. Fusion: The Energy Promise. NOVA Series. ¾" videocassette, 60 min., color. WGBH-TV, Boston, 1974. Dist.: Public Television Library. $150 for in-room use, consult catalog for rates for other uses; $70/2 weeks rental. Gr. 10–12.

Part of the NOVA series produced by WGBH-TV in Boston, this is the same program reviewed as a 16mm film distributed by Time-Life Multimedia (see No. 246).

248. Incident at Brown's Ferry. NOVA Series. ¾" videocassette or 16mm, 58 min. in 2-pts., color, w/tchr's. guide. WGBH-TV, Boston, 1977. Dist: Time-Life Multimedia. $520 videocassette; $650 16mm; $60/3 days rental. Gr. 10–12.

A film in the NOVA science series, this takes a long, serious look at the question of nuclear power plant safety in view of an accident started with a simple candle at the world's largest nuclear plant in Brown's Ferry, Alabama, in 1975. Dozens of nuclear scientists and engineers—citing government studies, computer analyses, and their own experiences in the industry—offer their views on the safety of the industry. Part I describes the Brown's Ferry accident and the frightening errors surrounding it which could have, but did not, result in a reactor meltdown. Conclusion: there was a failure to correct safety problems that had been identified. The 1975 Rasmussen report for the federal government is described and criticized for not anticipating the unexpected in this industry. Part II is somewhat drawn out. Critics of the Rasmussen study doubt the Atomic Energy Commission's credibility and Rasmussen's assumption that the safety systems are properly designed. The film contends that tests up to 1976 did not show with certainty that emergency cooling systems essential to reactor safety would work. Scientists at the National Engineering Laboratory in Idaho and at General Electric's reactor research facilities are interviewed and testing shown; one of the three GE nuclear engineers who quit the industry because of their concerns over nuclear design and safety states his position. Advocates say that the crucial questions are what safety margins are built in and what is the likelihood of an emergency, while those against nuclear energy raise the question "should we continue to live with a technology with such potential hazards?" and suggest increased pressure on the nuclear industry. Produced by WGBH in Boston, this film is for more advanced students in science and social studies. Some science teachers may feel it is better suited to social studies since the amount of scientific material included is small.

249. Introduction to Nuclear Fission and Fusion. 1 filmstrip w/1 cassette, 93 fr., color, w/tchr's. guide and 25 worksheets. Ward's Natural Science Establishment, 1976. $29. Gr. 10–12.

This semi-programmed study unit, intended for individual use, is divided into six sections with worksheet questions following each section. It describes the basic concepts of fission and, to a much lesser degree, of fusion, and pro-vides a very good explanation of critical mass and binding energy. It shows how heat is produced in a reactor by fission and points up the dead-end nature of fission use because of fuel shortages and the problem of radioactive wastes. The diagrams are not strong but are adequate for individual viewing. For senior high chemistry and physics students.

250. The Last Resort. 16mm, 60 min., color. Green Mountain Post Films, 1977. $625; $50 classroom and $75 auditorium rental for one showing, ea. additional showing half-price. Gr. 10-12.

On August 1, 1976, eighteen men and women were arrested for occupying the site of a planned nuclear power plant in Seabrook, New Hampshire, because they had been "outmoneyed and outlegaled" at a hearing on the plant and a negative town vote had been ruled illegal. Nine months later, the Clamshell Alliance of protestors took possession of the same site in a nonviolent attempt to halt construction. Using interviews, government film, on-site footage of the protest (with police removing the protestors), clips of the governor endorsing the plant, and folk and protest songs, the film's black and white and color footage examines the Seabrook confrontation as an example of civil disobedience and the worldwide conflict over the use of nuclear power. It is impressive despite its length.

251. Lovejoy's Nuclear War. 16mm, 60 min., color. Bullfrog Films, 1975. $600; $50/3 days rental. Gr. 10-12.

The story of a young New England farmer's efforts to alert his community to the dangers of nuclear power. The post-war development of nuclear power is recounted and the case is narrowed to Lovejoy vs. Northeast Utilities, a utility which planned to build a twin nuclear generating plant in the Mon-tague Plain area of western Massachusetts. Some citizens were concerned, others applauded the construction in an otherwise depressed area. In the film, Sam Lovejoy discusses the two actions open to citizens who objected to the plant—a hearing and a general intervention—and says he sabotaged a weather tower built by Northeast to predict radioactive stack emissions because he did not get a fair hearing and wanted to mobilize the com-munity. His arrest and the citizen reaction to it, the experts who came in to testify, and the power-company position on the construction of the plant are all covered in the film. Although his case was dismissed on a technicality, Lovejoy speculates that the court felt he was both sincere and right. The power of this film is mitigated by its length and by the fact that nowhere in it do either Lovejoy or his principal scientific supporter, Dr. John Gofman, get down to specific facts about the dangers from nuclear plants; instead, they repeat that the Atomic Energy Commission and Northeast did not give the facts to the people who must live with these plants. Some will find Lovejoy an attractive figure. The guitar strains of "It's Nice to Be Righteous, but Which Side Are You On?" are played against scenes of the lovely New England countryside. Most useful in social studies classes where examples of Americans acting against apathy and entrenched forces are under discussion.

252. The Need for Nuclear: Our Energy Options. videocassette or 16mm, 20 min., color. Atomic Industrial Forum, 1976. $100 videocassette; $150 16mm. Gr. 10–12.

Distributed by the trade association for the nuclear power industry, this studio production features Dr. Walter Meyer, Chairman of the Department of Nuclear Engineering at the University of Missouri. Meyer looks at the demand for power between 1972 and 2000 and the contribution each energy source can make in that period. He briefly examines the potential of oil, hydroelectric, geothermal energy, wind energy, solar energy, and shale oil, noting the problems of each, but crediting solar energy with the greatest potential value as the costs of other fuels rise and as large-scale production of solar systems lowers the unit price. Meyer concludes that in the short haul we will need both coal and nuclear power to move from the present resources to the alternate fuels of the future. He claims that nuclear power has not had a detrimental effect on health while coal—currently under study—may have. The presentation is reasonably objective considering the source. For discerning social science and science students.

253. Nuclear Energy. ¾" videocassette or 1" or ½" videotape, 28:50 min., color. Nebraska Educational Television Council for Higher Education, 1976. $225; $35/7 days rental. Gr. 10–12.

An introduction to the technical processes which make nuclear energy possible, this videotape begins with a discussion of binding energy and its function in holding atomic particles together and proceeds to explain the release of binding energy through either fission or fusion. The program concentrates on fission, describing what happens to uranium and plutonium isotopes which are struck by neutrons in a reactor and noting the function of the fuel rods. These processes are illustrated by photographs and animation of the subatomic processes. The conversion of the energy produced by the reaction into a usable power source is then discussed, and the two main types of reactors—the boiling water reactor (BWR) and pressurized water reactor (PWR)—are illustrated by schematic diagrams. Brief consideration is given to the environmental questions surrounding nuclear reactors, including the release of radiation into the environment, disposal of radioactive waste, and the lower thermal efficiency of nuclear plants. The orientation is pronuclear.

254. Nuclear Power: An Introduction. videocassette or 16mm, 18 min., color. Atomic Industrial Forum, 1975. $100 videocassette; $150 16mm. Gr. 10–12.

In this rapid-fire introduction to nuclear power, distributed by the trade association for the nuclear power industry, Dr. E. Linn Draper of the University of Texas at Austin describes steam-cycle production of electricity and the fission process and shows schematic diagrams of a reactor core and three types of reactors currently in use in the U.S. In addition, he describes the operation of the liquid metal fast breeder reactor expected to be operable in a demonstration plant by the middle 1980s. Almost half of the program is devoted to Draper's lecture on the multiple barriers built into nuclear power plants—part of a design philosophy he calls defense in depth—to insure safety. For good, discerning students.

255. Nuclear Power Reactor—Pickering. 10 slides, color, w/written narration. Donars Productions, n.d. $7. Gr. 10–12.

Produced by the National Film Board of Canada, this is a visual tour of the Pickering Generating Station, one of Canada's largest. One slide diagrams the reactor system; the remainder show the integral parts of this atomic power plant, as well as the "spent fuel bay," an indoor pool where uranium is stored after it has been used in the reactor. Although not visually exciting, the program provides some idea of how atomic power is generated. Some briefing on reactor design would be useful beforehand. The text accompanying these slides is printed in both English and French; requests to put this narrative on tape must be submitted to the producer for approval.

256. Nuclear Reactor Safety: An Overview. videocassette, 20 min., color. Atomic Industrial Forum, 1976. $100. Gr. 10–12.

Norman C. Rasmussen, a noted nuclear advocate from the Massachusetts Institute of Technology, lectures on nuclear reactor safety starting from the premise that America must depend on coal and uranium since no alternative technologies are developed enough to provide a major part of our energy supply. He explains that reactors cannot explode but that there can be a failure in the cooling system that allows fuel to melt and release radioactivity. However, he claims that the risk to U.S. citizens is one in five billion compared to one in 1,600 for accidents generally. Rasmussen stresses that "defense in depth," i.e., redundant back-up systems, has been built into nuclear power plants to anticipate engineering failures which cannot totally be prevented. He feels the dangers of sabotage have been exaggerated and does not consider nuclear power environmentally damaging to a society whose growing energy needs will not be met by attempts at conservation.

Oil

PRINT

257. Wade, Harlan, and Wrigley, Denis. **Oil.** Milwaukee, Wisc.: Raintree Children's Books, 1977. 32pp. $4.49 plb. Gr. K-1.

Imparting basic information clearly and effectively, this attractively designed picture book explains what oil does and where we get it.

258. Nixon, Hershell H., and Nixon, Joan Lowery. **Oil and Gas: From Fossils to Fuels.** A Let Me Read Book. New York: Harcourt, 1977. 64pp. $5.95 plb. Gr. 3-5.

Although the authors use the terms "gas" and "gasoline" in a confusing manner, they offer a clear explanation of how oil and gas are formed, how they are located, and, to a lesser extent, how they are used. Good charts and maps show U.S. and world oil and gas production. The book is supposed to be designed for new readers, but despite the large type, the use of unfamiliar terms makes it more suitable for intermediate or upper elementary students.

259. Neal, Harry Edward. **The Story of Offshore Oil.** New York: Messner, 1977. 64pp. $6.64 plb. Gr. 4-8.

The main purpose of this book is to show how oil is obtained by offshore drilling, focusing on the life of a crew on a Gulf of Mexico rig. The book does not flinch from discussing—and showing—the environmental objections to offshore drilling. A good glossary is included.

260. Ridpath, Ian, ed. **Man and Materials: Oil.** Reading, Mass.: Addison-Wesley, 1975. 33pp. $4.95 plb. Gr. 5-8.

This is a well-structured, informative book which uses photographs to explain what oil is, how it is located, drilling procedures, and how oil is transported and refined. The book includes particularly good diagrams of the fractioning and cracking processes, as well as a world distribution map. Although the vocabulary is somewhat difficult, the meanings of terms are made clear from the diagrams or defined in the text.

261. Doty, Roy. **Where Are You Going with That Oil?** Chicago Museum of Science and Industry Series. New York: Doubleday, 1976. 62pp. $5.95. Gr. 5-9.

The emphasis in this title is on the variety of ways oil is used in everyday life: for gasoline, diesel fuel, aviation, jet and rocket fuel, kerosene, industrial fuel oil, lubricating oils and greases, waxes (which turn up in candy, crayons and carbon paper, among other products), carbon black, coke, petrochemicals, plastics of all kinds, synthetic fabrics, automotive parts, building materials, cosmetics, furniture, medical uses, detergents and chemicals, and synthetic rubber. The self-contained spreads are packed with information and brightened by the author's detailed cartoons. The epilogue comes to grips with the hard facts of the oil shortage and suggests ways of conserving oil. An oil pipeline runs through the book, tying it all together.

262. Sullivan, George. **How Do They Find It?** Philadelphia, Penn.: Westminster, 1975. 160 pp. $7.50. Gr. 5-9.

Chapter four explains the "bright spot" technique of searching for oil on the continental shelves and slopes where as much as 190 billion barrels of oil and 1,100 billion cubic feet of natural gas may lie. It also describes the actual drilling processes. Useful for short science reports.

263. American Petroleum Institute. **Facts About Oil.** Washington, D.C.: American Petroleum Institute, 1977. 44pp. $.35; single copies free to teachers. Gr. 5–12.

The history of oil and how it is found, produced, transported, refined, marketed, and used are explained in short, heavily illustrated chapters. Sections on energy and the environment and conservation are cursory. A concluding chapter, "Energy and the Future," uses data from government, industrial, banking, and private research sources to show (with charts) high and low estimates of U.S. energy consumption, production, and imports through 1990. A one-page lesson plan based on this booklet and geared to middle or junior high school general science or social studies classes may also be obtained.

264. Foster, John T. **The Sea Miners.** New York: Hastings, 1975. 158pp. $6.95. Gr. 7–12.

In chapter four, Foster explains why he believes the world's energy needs will be met by mining the sea and describes the history and process of offshore drilling operations. Chapter nine, "Pollution," shows and describes the disastrous Santa Barbara oil spill and oil spills in other parts of the world which have had far-reaching effects on marine life and threaten the very existence of the seas. Useful in biology, oceanography, or ecology courses, as well as in general science and social studies classes.

265. Independent Petroleum Association of America. **The Oil Producing Industry in Your State.** Washington, D.C.: Independent Petroleum Assn., 1977. 108pp. free (30 copies maximum order). Gr. 7–12.

This reference tool offers statistical data on the nation's oil-producing industry, as well as breakdowns for each state of the extent and the economic value of oil production, reserves, exploration, and development. The last part of the book contains information of primary interest to the oil industry itself, though the table showing total petroleum imports by exporting country or area for the years 1957 through 1976 will interest many outside the industry as well. Sources for all statistics are given.

266. Independent Petroleum Association of America. **United States Petroleum Statistics 1977.** rev. ed. Washington, D.C.: Independent Petroleum Assn., 1977. $.10. Gr. 7–12.

This work includes national statistics from 1957 to 1976 on petroleum exploration, drilling, discoveries, proved reserves, oil supply, consumption, total imports, gas supply and price, composite value and cost of oil and gas, prices of crude oil and gasoline, and oil-industry costs.

267. Butterworth, W.E. **Black Gold: The Story of Oil.** New York: Four Winds Pr., 1975. 224pp. $6.95. Gr. 9–12.

This entertaining social and economic history of oil relates the growth of an industry which began with ten gallons a day at Oil Creek, Pennsylvania, and now boasts tankers which carry 2.2 million barrels each. Emphasis is on the disparate group of men who recognized the potential of oil and fought for control of the business in the 19th Century, turning it into a vast industry. Later chapters cover the part oil played in the development of the automobile, the Navy, and in the conduct of World War II. It also documents the notorious Torrey Canyon oil spill. Useful for reports on industrial development in America.

268. Roscow, James P. **800 Miles to Valdez: The Building of the Alaska Pipeline.** Englewood Cliffs, N.J.: Prentice-Hall, 1977. 227pp. $10. Gr. 9–12.

At a cost of nearly $8 billion and ten years in time, the pipeline may be, as the author says, "the largest and most expensive project ever undertaken by private industry." This is a year-by-year account of its building from the vantage point of the oil and construction companies involved. The author concludes that North Slope oil will help reduce American dependence on foreign oil more than any other action taken this decade. A *Booklist* reviewer called this "a sympathetic, congratulatory account of a phenomenal project that mentions, but does not confront, the diverse human, political, and technical issues involved."

269. Szulc, Tad. **The Energy Crisis.** rev. ed. New York: Watts, 1978. 152pp. $6.90 plb. Gr. 9–12.

Although brief chapters deal with the role of alternative energy sources in American life, the emphasis in this revision of Szulc's 1974 book is still on oil. Szulc blames the maneuvering and greed of the "American oil cartels, clumsily abetted by the United States government" for the energy crisis. He gives nearly equal credit to political administrations which supported or ignored the industry's machinations and failed to develop a national energy policy. Describing the economic and social effects of the energy crisis, he explains how the oil companies reaped enormous profits but inevitably lost a controlling voice in world oil economics. Szulc believes that mandatory disclosure of the energy resources data held by energy companies and active government involvement in domestic production, fuel imports, and fuel allocation is necessary to plan for the U.S. energy future. Lively but literate, this work provides the story behind the complex issue, and many students should find it intriguing.

270. Independent Petroleum Association of America. **America's Energy Dilemma: Facts versus Fallacies.** Washington, D.C.: Independent Petroleum Assn., 1977. 21pp. $.15. Gr. 10–12.

The industry's point of view on such "fallacies" as "the United States is running out of oil and natural gas" is presented here with clear red, white, and blue charts. Although the format may suggest a broad range of users, the reasoning and use of economic terms restricts it to social studies and economics classes in senior high schools.

271. Walton, Richard J. **The Power of Oil: Economic, Social, Political.** Somers, Conn.: Clarion/Seabury, 1977. 178pp. $7.95. Gr. 10–12.

In this tightly written history of American and world oil development, the author explores the complex economic, social, and political deals which have surrounded the industry since its beginning. The development-by-cartel of oil resources after World War I—when foreign companies made staggering profits in the Middle East and elsewhere—goes far toward explaining later retaliatory action by the Organization of Petroleum Exporting Countries (OPEC). Of special interest are the author's insights on oil pricing: "the price of oil for the consumer seldom has much to do with the costs of the producer and seller. Whatever the market will bear is what the oil industry will get, unless regulated by goverment." Walton notes the relaxation of environmental controls on oil production since the energy crisis and examines the prospects for conservation and discovery of new oil supplies. Useful in studying U.S. foreign policy, as well as domestic history.

272. Paust, Jordan J., and Blaustein, Albert P. **The Arab Oil Weapon.** Dobbs Ferry, N.Y.: Oceana, 1977. 370pp. $27.50. Gr. 11–12.

This compilation of papers, commentaries, speeches, articles, and legal references provides a detailed presentation of the legal and economic aspects of the Arab oil boycott and the United States's response to it. The book is divided into three sections: the oil embargo, the response to the embargo, and the legal framework. The introductory section gives an historical chronology of the embargo. Some of the papers, such as the statements and resolutions by the Organization of Petroleum Exporting Countries (OPEC), should be useful to students researching Arab-U.S. oil relations in courses on contemporary issues, international relations, or economics. The reprints of two Congressional "Committee Prints" titled "Oil Fields as Military Objectives: A Feasibility Study" and "Data and Analysis Concerning the Possibility of a U.S. Food Embargo as a Response to the Present Arab Oil Embargo" are of particular interest. Though the price will be too high for many school libraries, this work has reference value if there is need for it in the curriculum. Students will need a more general work on the topic before they can use this source.

NONPRINT

273. Fuel for Your Car. 1 silent filmstrip, 38 fr., color, w/tchr's. guide. Standard Projector and Equipment Co., Educational Projections Div., 1972. $7. Gr. 2–4.

The filmstrip shows how gasoline is obtained from petroleum, with the emphasis on the problems involved in drilling for oil. It provides a well-diagrammed description of how oil is obtained and processed. Self-testing frames are included throughout to check on and reinforce learning about the industry. The spelling frames included at the beginning and end of the strip tend to be a distraction, but on the whole the strip will be useful for either social studies or science units.

274. Oil: From Fossil to Flame. 16mm, 13 min., color, w/tchr's. guide. Centron Educational Films, 1976. $205. Gr. 5–8.

From Titusville to Alaska over a fairly standard but well-photographed route, this film shows where world oil reserves are found. It notes that despite reserves of 627 billion barrels, at our present rate of production and use these reserves will last only another 40 years. It discusses possible formation of oil and shows oil exploration with a geologist and geophysicist. Actual test drilling is described and diagrammed in fairly clear if rapid fashion, as is the process of fractionation at the refinery. The film ends with a look at the many products of the oil industry and concludes: "because it is so essential, oil will continue to play a major part in the economies of the industrialized nations."

275. New Oil for Energy. 1 silent filmstrip, 22 fr., b&w, w/tchr's. guide. Visual Education Consultants, 1973. $5.50. Gr. 5–9.

The 22 black and white frames very briefly describe where oil companies look for oil, how oil was formed, and what clues geologists look for in seeking oil. The strip is sketchy, although the good background material in the teacher's manual which can be read at eight points during the presentation will help to overcome this problem. Permission to record the content of the manual is given. Five key science words, with definitions, and three preview-review questions precede the visuals.

276. Offshore Oil Rig. 10 slides, color, w/written narration. Donars Productions, 1974, 1976 U.S. release. $7. Gr. 5–12.

With the current interest in offshore oil drilling, this fine-quality set produced by the National Film Board of Canada should be of interest to a wide range of students. It provides a guided tour of a $15 million "semi-submersible drilling platform" in the Atlantic Ocean. Although three slides show the actual drilling operation, the emphasis is on how the men live, how operations are monitored, and what safety measures are taken. The text accompanying these slides is printed in both English and French; requests to put this narrative on tape must be submitted to the producer for approval.

277. Oil. (Jackdaw). Comp. by Roger Vielvoye. 10 broadsides and booklets. Grossman Publishers, 1975. $5.95. Gr. 7–12.

A display-information pack containing ten broadsides and booklets which cover the following topics: how oil was found; its uses, transportation, exploration, conservation, and pioneers; and the "rise of the Sheikhs." It provides a mini-course or mini-display on the subject of oil. The kit, which was developed in England, is packaged in a plastic envelope.

278. The History of Oil: A Key to America's Growth (Set). 2 filmstrips w/1 cassette or 1 disc, 58–60 fr., 9:05-10:20 min., color, w/tchr's. guide. Multi-Media Productions, 1976. $16.95 set. Gr. 9–12.

In Part I, a retired railway conductor, Uncle Billy Smith, brings in America's first oil well, thereby changing America's home life and making it possible to operate industrial tools at higher speeds and to build larger, more powerful locomotives. The strip describes the rise of John D. Rockefeller, the development of the Los Angeles oil fields, how new uses for oil and oil products came into being, and the growing number of factories changing over from coal to water power to oil. Part II explains how oil became the ideal energy source for industry and the military, and the rapid integration of oil products into American society as the number of cars grew from 10,000 to 7 million in just 20 years. The beginning of co-operation between the oil industry and the federal government is traced from the 1930s through World War II. The program concludes by noting the ways in which contemporary America has come to depend on oil. Low in price, this is a refreshing, insightful social history which could be used in general science as well as social studies classes to draw attention to the problem of energy supply and demand.

279. Oil from the Earth. 1 filmstrip w/1 cassette, 80 fr., 20 min., color. RMI Media Productions, 1976. $30; $15 filmstrip; $15 cassette. Gr. 9-12.

This strip explains in detail what oil is, in what type of geological formations it is found (anticlines, faults, and stratigraphic traps), and how it exists within these formations. A section, "How It Is Found," demonstrates seismic readings: vibrations are sent into the ground and the travel time recorded and used to make contour maps. "How Drilling Is Done" describes who drills, how a well is completed, what techniques are used to increase oil recovery from the field, and the outlook for using oil to meet future energy needs. Oil from tar sands, oil shale, and coal gasification are briefly discussed. The strip is laudatory of the oil industry and is quick to explain the high costs incurred by the industry, since 89% of the wildcat wells are dry holes and only one out of 40 wells yields enough oil to cover the exploration costs. The program is detailed and its explanation of the processes by which oil is located, reached, and produced is particularly good.

280. Arctic Oil: The Trans-Alaska Pipeline. 16mm, 27 min., color. Professional Arts, 1977. $350; $45/3 days rental. Gr. 10-12.

When the trans-Alaskan oil pipeline became a reality in 1977, it represented the result of eight years of engineering and planning, the work of 70,000 people, and a cost of $9 billion. This film documents the technical problems which were surmounted to bring oil from the Prudhoe Bay area, 300 miles above the Arctic Circle, across the mainland of Alaska to the ice-free port of Valdez. The most interesting portion is probably the footage which shows how special shallow-draft tugs and barges were used to bring in materials needed to begin the operation during the brief six-week period when the ice in Prudhoe Bay breaks up. The process of assembling and shipping the massive equipment, the pipeline valued at $100 million, and the special modules used to create a processing plant for the crude oil are shown. Only passing reference is made to the "longest and loudest environmental debate in history" beyond the opening footage which shows how the danger to the permafrost lying beneath the tundra was surmounted. Mention is made of the production level of Bay fields (600,000 barrels daily at present, with 2 million barrels daily expected by 1980), but the film's real emphasis is on the enormous climatic and logistic obstacles which were overcome by American ingenuity (and a consortium of seven oil companies) in what it calls the largest privately financed project ever attempted. The film does not discuss how the oil obtained from these fields will contribute to our energy resources in the future and how long the oil supply will last. Although it could have offered more scientific and natural history data in relation to the pipeline and fewer shots of pipes and big machines, this film will probably be useful in American history or economics classes as a recent example of American inventiveness.

281. Tar Sands: Future Fuel. Energy Sources: A New Beginning Series. ¾" videocassette or 16mm, 27 min., color, w/tchr's. guide. Univ. of Colorado, Educational Media Center, 1975. $230 videocassette; $333 16mm; $2.50 tchr's. guide; $10/3 days 16mm rental; $10/3 days, $20/15 days, $30/30 days videocassette rental. Gr. 10–12.

One of the lesser known fossil fuels, found mainly in Alberta, Canada, and in Utah, tar sands yield an oil product called bitumen which can be processed and fractionated into such petroleum products as naptha, kerosene, and gas oil. This film describes in traditional instructional style the two methods of producing oil from sands: surface or open-pit mining and thermal recovery. The former requires huge equipment and extensive manpower to replace and reclaim the excavated sands. Thermal recovery involves *in situ* flooding and forward or reverse combustion to release the oil deep underground. Because the development of this resource is hampered by the lack of a policy allowing sands leases on federal lands in Utah and by the high cost of technology, the film implies that tar sands are at best an interim answer to the energy crisis.

282. Oil Shale: The Rock That Burns. Energy Sources: A New Beginning Series. ¾" videocassette or 16mm, 29 min., color, w/tchr's. guide. Univ. of Colorado, Educational Media Center, 1975. $230 videocassette; $333 16mm; $2.50 tchr's. guide; $10/3 days 16mm rental; $10/3 days, $30/30 days videocassette rental. Gr. 11–12.

Most of this film is devoted to explaining precisely what shale is and the two processes—above-ground retorting and *in situ* retorting—which can extract 33 gallons of oil from a ton of shale. It shows the location of our principal shale reserves in the Green River Formation in Utah-Wyoming-Colorado, where the Department of the Interior has leased the rights to public lands for private exploration, and describes the technology involved in separating oil from the shale. Although *in situ* retorting is described as the more economical and desirable method since above-ground retorting leaves producers with spent shale and problems of revegetation, the film notes the strong citizen objections to *in situ* retorting when nuclear devices are used for blasting. This film appears to be comprehensive with one exception: its analysis of the relationship of shale oil to present and future energy shortages is confusing and inconclusive.

Pedal Power

PRINT

283. McCullagh, James C., ed. **Pedal Power in Work, Leisure and Transportation.** Emmaus, Penn.: Rodale Pr., 1977. 133pp. $4.95 pap. Gr. 10–12.

A reflection of the renewed interest in pedal power as a low-technology energy device for both developing and developed countries, this work suggests scores of tasks which can be accomplished easily and effectively with pedal devices. Chapters cover the use of human muscle power in history and contemporary uses of pedal power in Third World countries. One chapter provides detailed instructions for building an experimental "Energy Cycle," a multi-use foot-powered generator. Sources of equipment and additional information are included in the appendix. This is an attractive book which will appeal to tinkerers in and out of the classroom.

NONPRINT

284. Pedal Power. 16mm, 18 min., color. Bullfrog Films, 1978. $250; $28.50/3 days rental. Gr. 7–12.

Leonardo, who understood human physiology, designed a lathe powered by a treadle in 1500. At the end of the 19th Century, the internal combustion machine which operated on fuel resources edged out pedal and treadle machines. Today, with the search for more efficient, nonfossil energy sources, the strength in human legs is being rediscovered and pedal power is offered as the supreme example of technology on a human scale. Stuart Wilson of Oxford University's Engineering Lab explains the advantages of mechanical power over muscle power. The film goes on to demonstrate ways of putting pedal power to work. Richard Ott, a designer who has adapted a bicycle to tasks such as grinding grain or plowing a field, claims 95% efficiency for his "Energy Cycle." The film stylishly applauds the bicycle, shows ways the "Energy Cycle" is used, and asks "Do we really need nuclear power to boil an egg?" A trailer notes that complete instructions for building an "Energy Cycle" may be found in Rodale Press's book, *Pedal Power* (see No. 283).

Reference

PRINT

285. World Book Encyclopedia. **The Energy Problem.** Chicago, Ill.: World Book-Childcraft International, 1976. 8pp. 1–19 copies, $.20 ea.; 20 or more copies, $.15 ea. Gr. 4–7.

This illustrated learning guide explains how the *World Book Encyclopedia* can be used as an information resource on energy. The guide lists relevant *World Book* articles and raises pertinent study questions under such topics as "Early Power Sources: Muscle, Wind, and Water" and "Energy Sources of the Future."

286. Sobel, Lester A., ed. **Energy Crisis.** 3 vols. New York: Facts on File, 1974, 1975, and 1977. Vol. 1, 1969–73, 262pp., $11.95; Vol. 2, 1974–75, 213pp., $10.95; Vol. 3, 1975–77, 223pp., $11.95. Gr. 5–12.

These three reference tools, based on the weekly news coverage of *Facts on File*, record developments in the world energy situation from 1969 through mid-1977. In volume one, the approach is chronological and then by topic; the topics vary each year. Volumes two and three take a topical approach. For example, volume three is divided into sections called "U.S. Policy," "Resources and Tactics," "International Developments," "Worldwide Domestic Developments," and "Nuclear Energy." Users may find the tables of contents as helpful as the indices.

287. Sobel, Lester A., ed. **Jobs, Money and Pollution.** New York: Facts on File, 1977. 216pp. $10.95. Gr. 9–12.

This reference book provides a chronology of pollution events and steps to mitigate pollution during the period 1969–1977. It is based on the weekly *Facts on File* reports. Each of five sections recounts happenings regarding pollution, some of which—e.g. auto pollution, thermal pollution, the nuclear issue, oil pollution—are energy-related. For topical reports, both the table of contents and the index must be consulted. The index is not totally adequate; there is a heavy emphasis on proper names, terms that deserve a separate entry are buried under general entries, and several items mentioned in the text could not be located in the index.

288. American Geological Institute. **Dictionary of Geological Terms.** Garden City, N.Y.: Anchor Pr., 1976. 472pp. $3.50 pap. Gr. 10–12.

Published in 1957 and 1960 under the title *Glossary of Geology and Related Sciences*, this paperback was compiled by more than 90 scientists under the direction of the AGI and defines nearly 8500 entries covering all the major disciplines of geology. Generally technical, it can be useful to science classes discussing terms related to the production of various forms of energy.

289. Lapedes, Daniel N., ed. **McGraw-Hill Encyclopedia of Energy.** New York: McGraw-Hill, 1976. 795pp. $24.50. Gr. 10–12.

This reference work contains more than 300 articles written by specialists on the economic, political, environmental, and technological aspects of energy. The first section, "Energy Perspectives," features articles on energy consumption, the outlook for fuel reserves, exploring energy choices, U.S. policies and politics, protecting the environment, and the world energy economy. In the second section, "Energy and Technology," there are 300 alphabetically-arranged articles of varying length and complexity. Although the writing is clear and objective, many of these articles require considerable background in science. Coverage is given many subjects which rarely receive comprehensive treatment elsewhere, including laser-induced fusion, magnetohydrodynamics, nuclear fuels reprocessing, nuclear fusion, the heat pump, and hydrogen-fueled technology. The encyclopedia is extensively illustrated with photographs, charts, diagrams, etc. This is not a source for general science assignments, but it could be useful for more advanced students doing science research projects at the upper secondary level.

290. Public Affairs Clearinghouse. **Energy: A Guide to Organizations and Information Resources in the United States.** 2nd ed. Claremont, Calif.: Public Affairs Clearinghouse, in prep. $20. Gr. 10–12.

This edition of *Energy* was not available for evaluation, but it was expected to be released in April, 1978. The publisher states that it describes "more than 1500 public and private organizations by subject, including full address and telephone number, names of key officials, structure, programs, activities and publications . . . Information is fully indexed by name of organization, key words, subject, acronyms and initialisms . . . For easy reference the book is organized into ten chapters by major topic: energy in general, oil and gas, coal, water power, nuclear fission, alternate sources of energy, electric utilities, energy conservation, environmental impacts, and consumer aspects." The first edition of this book was chosen by *Library Journal* as one of the "Best Sci-Tech Books" of the year.

Solar Energy

PRINT

291. Berger, Melvin. **Energy from the Sun.** Let's-Read-and-Find-Out Science Book. New York: Crowell, 1976. 34pp. $5.79 plb. Gr. 2–4.

A new-reader science book of remarkable completeness, this describes what energy does and how we get it: either directly through the food cycle or indirectly by using machines and electricity which depend on fossil fuels for their supply of energy. The emphasis is on the sun as the source of all energy. Giulio Maestro's bold drawings in black and white and hot colors add interest.

292. Asimov, Isaac. **What Makes the Sun Shine?** Illus. by Marc Brown. Boston, Mass.: Atlantic Monthly Pr., 1971. 57pp. $4.95 plb. Gr. 5–8.

Asimov explains how the sun and planets were formed, the many forms of energy, and the fusion reaction which takes place in the sun between hydrogen and helium to produce the huge amounts of surplus energy—radiation—we see as light and feel as heat. A good glossary with page references is included. Although the format is geared to the young, the vocabulary and concepts are suitable for the middle school and up.

293. Branley, Franklyn M. **Solar Energy.** New York: Crowell, 1975. 117pp. $5.95. Gr. 5–9.

Characterizing the sun as a "vast, almost everlasting source of energy that has never been effectively utilized," Branley, Astronomer Emeritus of the Hayden Planetarium, explains where solar energy comes from, how a solar heater works, the principle of a heat pump, and solar space heating. He is concerned with the relative inefficiency of the sun in producing food and suggests algae farms be cultivated as an efficient solution to the world food problem. Solar furnaces, steam boilers, cookers, water distillers, and generators are also discussed. Clear, simple diagrams showing how each system works are provided, but there is no discussion of the economic practicality of these systems today.

294. Knight, David C. **Harnessing the Sun: The Story of Solar Energy.** New York: Morrow, 1976. 128pp. $5.95; $5.49 plb. Gr. 5–9.

Sections in this book present a good history of man's experiments with solar energy, describe the ways solar energy can be collected, show and explain how solar energy is used to heat water and buildings, and explore the possibilities of the sun as a source of electric power. Thoroughly researched but somewhat didactic in its writing style, this will be useful for science reports at the upper elementary and junior high level, or with slower general science classes in secondary schools.

295. Halacy, D. S., Jr. **Solar Science Projects for a Cleaner Environment.** New York: Scholastic, 1972. 96pp. $.95 pap. Gr. 7–12.

Formerly titled *Fun with the Sun,* this low-cost paperback describes how to build seven small-scale solar energy projects, including a reflector cooker, a solar still, a furnace, an oven, a water heater, a motor, and a radio. While the directions are clear enough, obtaining the materials and assembling the projects will require considerable assistance from the teacher for most junior high science students. For independent use, it is better suited to vocational education classes (including first-year electronics programs) or to handy senior high science students.

296. Gunn, Anita. **A Citizen's Handbook on Solar Energy.** Washington, D.C.: Public Interest Research Group, 1976. 50pp. $2 to individuals; $10 to institutions. Gr. 9-12.

The publisher of this manual is an organization maintained by Ralph Nader to perform research on public policy issues, including energy. It is not a true handbook but a layman's introduction to different solar technologies, e.g. bioconversion, solar thermal conversion, photovoltaic energy. Ideas such as decentralization of energy control through solar energy, and solar funding and nonfunding, are also discussed. A useful list of experts and information sources on solar energy, which notes their areas of expertise, is included. PIRG also offers an annotated list of short publications on energy, with special emphasis on nuclear energy; titles include "Oil Pricing Myths Invented by the Oil Industry," "What's Wrong with the Atomic Industry?," and copies of testimony before various governmental committees and agencies.

297. Halacy, D. S., Jr. **The Coming Age of Solar Energy.** rev. ed. New York: Harper and Row, 1973. 231pp. $9.95; Avon, $1.95 pap. Gr. 9-12.

Halacy provides a history of solar energy research throughout the world from Archimedes to the early 1970s, and he attributes the revival of interest in solar energy to space exploration. Practical applications of solar energy are given, but more space is devoted to photochemistry and to electric power from the sun. Chapters are directed to proposals for an orbiting solar power plant, a vast National Solar Power Plant Facility in the Arizona-California desert, and to the utilization of ocean thermal energy. Nontechnical, informative, but somewhat pedantic, this work is recommended for science buffs and senior high science or social studies classes.

298. Keyes, John. **The Solar Conspiracy.** Dobbs Ferry, N.Y.: Morgan and Morgan, 1975. 124pp. $3.95 pap. Gr. 9-12.

The author, whose research and development group invented a functioning backyard solar furnace, contends that there are no technical impediments to the introduction of solar heating today. He questions the research program which concentrates on high-technology centralized solar installations and costly water-based solar heating for homes, stating that this program has grown without proper coordination or the goal of designing practical, inexpensive solar systems. Keyes fears that Americans may end up "paying for the sun" if solar energy is centrally converted and solar installations are leased or rented rather than purchased outright by homeowners. A readable, action-oriented book which decries government ineptness and/or collusion and big-business interference in a field which Keyes feels should be a small-business operation. Appendices contain useful advice on conservation and sections for the homeowner, builder, and architect considering solar heating. An allegorical tale precedes each chapter.

299. Rankins, William H., III, and Wilson, David A. **Practical Sun Power.** Black Mountain, N.C.: Lorien House, 1974. 52pp. $4 pap. Gr. 9–12.

This photo-offset, hand-assembled book describes five basic solar projects which can be completed without any special tools or hard-to-get materials. The projects include parabolic reflectors, a solar oven, a hot water heater, a window air heater to supplement a regular heating system, and thermocouples. It also makes suggestions for mini-projects and other solar experiments. Each project contains clear step-by-step building directions, advice on scrounging materials, and an occasional word of caution. ("If you decide to construct a *huge* reflector, assemble it at night, or under a tent. If done in the daytime, you may neatly roast *yourself.*")

300. Alves, Ronald, and Milligan, Charles. **Living with Energy.** New York: Penguin, 1978. 128pp. $5.95. Gr. 10–12.

This attractively illustrated color and black-and-white picture book is primarily a round-up of variously designed solar homes in several American locations. It also describes greenhouses, a community recreation center, and wind power systems. "Whole-life systems" such as the Integral Urban House in Berkeley, the New Alchemy Institute on Cape Cod, the Institute for Social Ecology at Goddard College in Vermont, and the Earthmind headquarters in California show energy pioneers "learning to live in greater harmony with the renewable resources at hand." The authors assert that well-engineered alternate energy systems are available and cost-competitive now. Though they applaud the recent increase in federal solar research spending, they are concerned that the large scale, high technology tack this research is taking discourages regional or individual approaches. A glossary and extended list of sources for hardware and information add to this title's usefulness for teachers and students.

301. Cassiday, Bruce. **The Complete Solar House.** New York: Dodd, 1977. 212pp. $8.95. Gr. 10–12.

This is an attractive, comprehensive guide to the possible applications of solar energy to home heating. It provides background on solar heating and in separate chapters describes many examples of solar hot-water systems, solar-heated swimming pools, and new and retrofitted homes using solar space-heating and cooling. References to experimental homes and solar research are current. Concluding chapters discuss the cost of solar heat, zones of insolation (the total amount of solar radiation received), and, briefly, electric power from the sun. Manufacturers of solar components are listed. Useful as a reference in science, industrial arts, or building trades classes for students with some technical background.

302. Ewers, William L. **Solar Energy: A Biased Guide.** Northbrook, Ill.: Domus Books, 1977. 96pp. $4.95 pap. Gr. 10–12.

Frankly pro-solar, this well-designed, sometimes technical work combines the history of solar energy development and speculations about its future development with descriptions of practical applications of solar energy such as water heating, space heating and cooling, and swimming pool heating. It also describes public buildings already using solar energy. The author calls for government involvement to insure the adoption of solar power as an energy source. There is a good current bibliography and a list of suppliers for solar systems or components.

303. Fleck, Paul A., ed. **Solar Energy Handbook.** rev. ed. Pasadena, Calif.: Time-Wise, 1975. 90pp. $3.95. Gr. 10–12.

A mixed bag of solar information with an emphasis on technical aspects of solar energy use, this handbook gives energy units, radiation units, and power units useful for calculations in building solar collectors, along with definitions and abbreviations of solar-related terms. Also included are figures on energy reserves, power consumption, and energy production and capacity through 1974, and a 20-page summary on the uses of solar energy.

304. Hayes, Denis. **Energy: The Solar Prospect.** Worldwatch Paper No. 11. Washington, D.C.: Worldwatch Institute, 1977. 80pp. $2 pap. Gr. 10–12.

This is a succinct, highly readable paper. Hayes believes that it will take a combination of solar technologies to meet the world's energy demands. He describes these technologies under such broad headings as: solar heating and cooling, electricity from the sun, catching the wind, falling water, plant power, and storing sunlight. Each technological system is explained in terms of where it can be used and why it is important. Hayes is optimistic about the prospects for the photovoltaic cell, noting that increased purchases have caused a drop in their prices. He is realistic about the need for energy-intensive industries to relocate in sunny climates since long-distance transmission of electricity will continue to be expensive. Advocating a new energy era based on small-scale decentralized solar technologies, he believes that the transition to the solar age can be begun today. Although there are no illustrations and no index, this work contains a great deal of information which will be useful to the layperson. This book is adapted from Hayes's *Rays of Hope: The Transition to a Post-Petroleum World* (see No. 175).

305. Morris, David. **The Dawning of Solar Cells.** Washington, D.C.: Institute for Local Self-Reliance, 1975. 16pp. $2. Gr. 10–12.

This working paper explains that solar cells, devices which convert sunlight to electricity, can supply almost 100% of our electrical requirements by the year 2000 and significant amounts of electricity within the next decade. It notes, however, that solar cells have few economies of scale, so solar-cell technology will require reevaluation of our present system of centralized and regional generating plants. The author sees a need for an artificial market to be created for a short period so that the nation may be brought to self-sufficiency in electric-power generation. This book includes a bibliography and list of manufacturers. Somewhat technical by definition, it could be used in economics and contemporary issues groups, as well as in physics classes.

306. Anderson, Bruce, and Riordan, Michael. **The Solar Home Book: Heating, Cooling and Designing with the Sun.** Harrisville, N.H.: Cheshire Books, 1976. 297pp. $8.50 pap. Gr. 11–12.

The student who is curious about solar energy will be interested in the descriptions of early and recent solar buildings, the explanation of such "soft technology" approaches as water walls and roof ponds, and the detailed information on heat collection, distribution, and storage provided in this book. Written by an architect, the book is more a browser's introduction to solar's uses in the home than a how-to guide.

307. Baer, Steve. **Sunspots: Collected Facts and Solar Fiction.** Albuquerque, N.M.: Zomeworks Corp., 1977. 146pp. $4. Gr. 11–12.

An off-beat, somewhat choppy collection of anecdotes, fantasies, facts, and projects connected with solar energy, this has useful information on solar heat collectors, insulating devices, and storage systems. The publishers build zomes and solar heating systems and have a consulting-designing service for architects who want to include energy-conserving devices in new designs or existing buildings.

308. Build Your Own Solar Water Heater. Winter Park, Fla.: Environmental Information Center of the Florida Conservation Foundation, 1976. 25pp. $2.50. Gr. 11–12.

Solar water heaters have been used in Florida since before 1900; now, with rising fuel prices, they are making a comeback. This practical booklet tells how to build a solar heater with materials generally available to vocational education classes. There are many good illustrations, lists of materials, and step-by-step instructions.

309. Daniels, Farrington. **Direct Use of the Sun's Energy.** New Haven, Conn.: Yale Univ. Pr., 1964. 271pp. $15. Ballantine, $1.95 pap. Gr. 11–12.

This title, considered a "classic" in the solar energy field, is for the student with a strong interest in solar energy. Although its purpose is to interest scientists and engineers in undertaking research on the direct use of the sun's energy, it could be used by the student who is looking for in-depth information on the many avenues solar research has taken: from agricultural and industrial drying to photovoltaic conversion. References made are generally to scientific and technical journals.

310. Daniels, George. **Solar Homes and Sun Heating.** New York: Harper and Row, 1976. 178pp. $9.95. Gr. 11–12.

A book for the layman, this explains how solar heating works, how to free heat, how to hold on to stored heat, how to build a flat-plate solar heating system, and how to bring solar heat to an existing house. Photographs and schematics are used. Technical without being obscure, it could be read in heating/air-conditioning and architecture courses. The author is a former editor of *Science Illustrated* and *Mechanix Illustrated.*

311. Hudson Home Guides. **Practical Guide to Solar Homes.** New York: Bantam, Hudson Home Publications, 1977. 143pp. $6.95 pap. Gr. 11–12.

A reference book for the carpentry, heating/air-conditioning, or architecture student who wants to know about solar system components, heat storage, and energy distribution. The book shows different types of solar heating systems and many handsome examples of solar homes across the country and includes 60 pages of renderings and floor plans for solar or energy-conserving homes. A section called "The Solar Shopper" gives examples of solar components offered by manufacturers.

312. Lucas, Ted. **How to Build a Solar Heater: A Complete Guide to Building and Buying Solar Panels, Water Heaters, Pool Heaters, Barbecues and Power Plants.** Pasadena, Calif.: Ward Ritchie, 1975. 236pp. $4.95. Mentor, $2.25 pap. Gr. 11–12.

This guide provides directions in various levels of detail for building solar collector panels, solar heating systems, a solar water heater, a swimming pool heater, and solar cooking equipment. The last chapters merely describe solar cells and backyard and large-scale power plants. The book will be helpful for students in heating/air-conditioning or carpentry classes, but some background in electricity and plumbing will be needed.

313. Solar Energy Handbook: Special California Edition. Pasadena, Calif.: Time-Wise, 1977. unp. $5.95 pap. Gr. 11–12.

This book was written in response to the Hart Law which provides a state income tax write-off of up to $3000 to California homeowners who install solar energy systems. It shows Californians how to plan a space or water heating system using solar energy, how to install the system, and how to claim tax credit. The clear writing and step-by-step instructions make this work a helpful resource for heating and air-conditioning classes in California. The text of the Hart Law and sections of the *Solar Energy Handbook* are included in the appendices.

314. Wade, Alex. **30 Energy-Efficient Houses You Can Build.** Photos. by Neal Ewenstein. Emmaus, Penn.: Rodale Pr., 1977. 316pp. $10.95; $8.95 pap. Gr. 11–12.

An architect for 20 years, Wade describes the rationale, the design, and the cost of 30 small houses which combine aesthetics with space and energy economy. Chapters on south-facing hillside houses and passive solar-tempered homes are particularly relevant. Many attractive black and white photographs and drawn-to-scale plans make this a good source book for students in carpentry or construction courses.

315. Wells, Malcolm, and Spetgang, Irwin. **How to Buy Solar Heating without Getting Burnt!** Emmaus, Penn.: Rodale Pr., 1978. 262pp. $6.95 pap. Gr. 11–12.

Although this title is directed at home owners, much of the information in it can be used by students in architecture and air-conditioning/heating classes who are interested in solar energy. Chapters cover insulation, deter-mining the suitability of houses for solar applications, dealing with contracts and contractors, and legal and financial considerations. An appendix lists well-established manufacturers of solar collectors and solar kits. Chapter five lists manufacturers of solar heating equipment.

316. Foster, William N. **Homeowner's Guide to Solar Heating and Cooling.** Blue Ridge Summit, Penn.: Tab Books, 1976. 196pp. $7.95; $4.95 pap. Gr. 12.

This technically oriented paperback describes solar heating systems, system components, typical system arrangements, and the size of systems. This title also offers a practical opening chapter on the economic feasibility of solar heating compared with other energy sources. It concludes by offering advice to the potential consumer on purchases and legal problems which should be considered. A glossary and list of solar manufacturers are included.

NONPRINT

317. Our Energy Source: The Sun. 1 filmstrip w/1 cassette, 41 fr., 7:58 min., color, w/tchr's. guide. Multi-Media Productions, 1976. $14.95. Gr. 4–8.

This strip provides brief, basic information on the sun as the source of all the world's energy, its necessity for plant and animal life, and its part in photosynthesis and in determining the world's weather.

318. Solar Energy (Set). 2 filmstrips w/2 cassettes or 2 discs. 51–59 fr., 12–15 min., color, w/tchr's. guide. Educational Activities, 1977. $27 set w/discs; $29 set w/cassettes. Gr. 4–8.

Part I, *The Sun: An Old Solution to a New Problem*, traces the use of the sun throughout history and shows applications of solar energy today: solar panels, solar cells on satellites, a solar power plant using giant mirrors. Although this is a once-over-lightly presentation, it does get across important facts: that the sun's energy is renewable, clean, inexhaustible, pollution free, and distributed equitably around the world. Part II, *Putting the Sun to Work*, goes into detail about the two main problems inherent in solar energy utilization: collection and storage of energy. The principle of black heat-absorbent surfaces is explained and collecting devices using water and air are shown. Methods of storing solar energy so it can be used in the evening or on sunless days are presented. The heat pump system of home heating and cooling, solar concentrators for power plants and furnaces, thermal power plants in the ocean, and solar cells are all discussed. This set is useful for its exposition of the principles behind utilization of solar energy; it does not provide a thorough analysis of the technical difficulties or the costs involved in using solar energy as opposed to fossil fuels.

319. The Sun: Its Power and Promise. 16mm, 24 min., color, w/tchr's. guide. Encyclopaedia Britannica Educational Corp., 1976. $320; $17 rental. Gr. 5–8.

Worshipped by ancient man, the sun today represents light, heat, and the promise of continued life, as well as a new source of energy for the future. Through astrophotographs, this film looks at the sun itself, its spots and prominences, and discusses its potential for uses such as cooking, high temperature furnaces, and solar heating. Somewhat more offbeat uses such as solar watches, radios, and calculators are shown. The scientific content of the film is negligible, but it could introduce the idea of solar energy to younger students.

320. Switch on the Sun. rev. ed. 16mm, 15 min., color, w/tchr's. guide. Xerox Films, 1977. $255. Gr. 5–8.

The sun was worshipped by ancient peoples but ignored by modern America as long as fossil fuels were cheap and plentiful. This film presents a short overview of the sun's possibilities as an energy source. The secret is to capture, concentrate, and convert its energy. Examples of applications of solar energy are shown: solar collectors, solar cells, solar homes in California and New Mexico, an experimental school in England which is partly solar heated, and the French solar furnace. Solar energy is seen as a likely source for poor nations if it is developed on a large scale and can provide ready energy inexpensively. A light treatment, the film uses some of the same footage as Xerox's *Power without End* (see No. 61).

321. Solar Energy. 1 filmstrip w/1 cassette or 1 disc, 59 fr., 9:55 min., color, w/tchr's. guide. Random House Educational Media, 1977. $19.50. Gr. 6–9.

Brief but comprehensive, this introductory strip presents the principal ways of utilizing solar energy—by collectors, warming air or warming water; by concentrating the sun with huge mirrors; or directly, by means of solar cells—and the problems involved with each. It goes on to discuss the main areas which offer hope of using solar power to its fullest capacity: a solar satellite or power station using microwaves to beam solar-cell energy to earth and "bringing the sun down to earth" in a fusion reactor. Good, clear, ungimmicky explanations for junior high science students.

322. Solar Energy (Set). 2 filmstrips w/2 cassettes, 50–80 fr., 8–14 min., color, w/tchr's. guide. Educational Dimensions Group, 1976. $39 plus $.50 postage/handling. Gr. 6–10.

A nontechnical introduction to the possibilities of solar energy. Part I shows how solar energy is received and stored, using beaches and rocks as examples, and describes the applications of solar power in home heating. It is somewhat downbeat about the immediate prospects for solar heating and is given to unsupported generalities such as "the cost of installation is more than most people can afford" and "many people don't want such monstrosities in their homes" (referring to storage sections), before concluding that solar technology problems *can* be solved. Part II shows several new applications of solar energy: two methods of heating swimming pools, solar greenhouses which increase the growing season and the food supply, solar "fish farms" which offset pollution and over-fishing in the oceans, an experimental solar furnace which uses a heliostat to reach 5000 degrees to burn a hole through aluminum, and a solar space station. All of these do not get equal coverage. Although the technical problems involved in these solar applications are not discussed, this is a visually interesting presentation of some seldom-discussed uses of solar energy.

323. The Solar Frontier. 16mm, 25 min., color. Bullfrog Films, 1977. $350; $33.50/3 days rental. Gr. 7-12.

This beautiful film combines nature and technology to explore the seldom-raised question, can solar energy be used to heat homes where it matters most, in the snowbelt during the long winters? Three Canadian solar homes—one of them in a suburban subdivision—are shown, and their owners talk about the economies and pleasures of living in them. The architects who designed the homes discuss the importance of an energy-efficient design and excellent insulation to prevent the loss of captured heat. The type of solar system varies in each house; each is diagrammed and overall costs are quoted. The film is effective because it deals not with the theoretical but with average people living in real homes. Its one flaw might be that no problem in any of these heating systems is suggested, presenting a rather roseate view to an untutored audience. Science teachers will need to do homework on solar heating to present the technical problems as well as advantages.

324. Desert Cloud. 16mm, 8 min., color. Bullfrog Films, 1976. $250; $25/3 rental. Gr. 9-12.

A visual tour de force about a plastic "solar air structure" whose architect designer has used the principle of black-body radiation to inflate the structure and condense water, which can then be used for agricultural irrigation. Other applications are suggested, but the emphasis is on the structure itself as a solar device utilizing basic physical principles for practical purposes. No indication is given as to whether the prototype has been put into production. Set against Middle Eastern desert scenery, the solar "raft" undulating sinuously in the sky, this is a handsome film. Students, however, may find the English narrator, who is sometimes overshadowed by the Kings College, Cambridge University choir singing celestial music, difficult to follow. Stressing the "ambience" between the sun and other natural phenomena, this could be used for enrichment in physics classes with students who already understand radiation principles. Younger, less knowledgeable students may feel it is a film about Kuwait.

325. How to Make a Solar Heater. 16mm, 20 min., color, w/tchr's. guide. Handel Film Corp., 1977. $290; $29/7 days rental. Gr. 9-12.

A well-constructed, uncomplicated film which shows how Jim, a teenage boy who wants to heat his workshop, goes about designing and building a solar heating system. Aided by a visit to NASA's Jet Propulsion Laboratory to learn how large his solar collecting system must be, he scrounges and buys simple materials—among them, a garden hose, tubing, wood framing, plastic, and a 55-gallon drum—for less than $100 and proceeds to build his collection system. The narrator describes what Jim is doing at each step and how his closed-circuit heating system works. Useful for general science, industrial arts, or building trades projects to show a practical application of solar energy.

326. Solar Energy (Series). 6 ¾" videocassettes, 30 min. ea., color. KNME-TV, Albuquerque, 1975. Dist.: Public Television Library, Video Program Service. $130 ea. for in-room use, check catalog for rates for other uses; $34 ea./2 weeks rental. Gr. 9–12.

Produced by KNME-TV in Albuquerque, the series examines crucial questions about solar energy: how much of the sun's power can be harnessed, how soon can solar technology ease our dependence on petroleum reserves, and how much is it going to cost? The six programs are: *Phase Zero; The Theory Is Tested; The Do-It-Yourself Guide to Solar Living; Power; The Solar Scenario;* and *The Solar Decision.* Only two of the six were available for preview. Both contain provocative information and footage not seen elsewhere in nonprint materials on solar energy. *Phase Zero* considers two applications of solar power: heating and electric power. It visits a Colorado architect who says his solar heating installation provided 70% of the heat for his home the first winter. It also shows the first solar subdivision in the U.S.—in California—and provides examples of state legislation which encourage solar housing. The narrator prophesies that by 1985 one out of ten new homes will be solar-assisted. The second half of the program considers how solar energy can be converted to electric power, examines Sandia Laboratories' wind power experiments in New Mexico, and tours a Martin Marietta Aerospace research installation which converts sunlight to steam to run turbines. Solutions to solar shortcomings are explored, as are a solar cell satellite and a "solar energy plantation" on which trees will be raised, dried, and burned as fuel for electric generating plants. The program estimates that by the 21st Century 20% of our energy needs will be met through solar power. In *Power,* some of the research facilities visited in the first film are examined in more detail. The concept of total energy systems is being considered by Sandia Laboratories, who propose to use solar energy left over from generating power to heat homes and water. The possibility of wind power utilization in coastal areas and in the Great Lakes and Plains areas is considered, as is a "digestive system" using animal waste from the world's largest cattle feedlot to produce methane for electric power. The narrator concludes that federal research monies for solar energy research are still so small—$50 million in 1975, the cost of one medium-sized oil tanker—that it is no wonder that the cost of using solar energy is not yet competitive with that of fossil fuels.

327. Applications of Solar Energy. ¾" videocassette or 1" or ½" videotape, 28:58 min., color. Nebraska Educational Television Council for Higher Education, 1976. $225; $35/7 days rental. Gr. 10–12.

This educational-TV production concentrates on the use of solar energy for heating and cooling buildings, using a demonstration home near Lincoln, Nebraska, as a functioning example. Richard Bourne, professor of construction management at the University of Nebraska and design consultant for the house, narrates the program. He discusses active and passive solar heating systems and their engineering efficiency, comparing the costs of operating solar and nonsolar heating and cooling systems in areas with different fuel costs. Using diagrams and slides of the actual house, he describes in detail a home using both solar heat and a heat pump, concentrating on the cost-effective qualities of this house. Good diagrams explaining solar heating systems make this useful in building trades and general science classes.

328. Dawn of the Solar Age, Pts. I and II. NOVA Series. ¾" video-cassette or 16mm, 28–29 min., color. Time-Life Multimedia, 1977. $520 videocassette; $650 16mm; $60/3 days rental. Gr. 10–12.

Far more detailed than other films on the subject, Part I of this NOVA science film, *Solar Energy*, takes up the problems entailed in using clean, everlasting solar energy and doing it cheaply enough to compete with or displace nuclear power. Several methods of collecting and converting solar energy to electricity are shown and their attendant problems are brought out. For example, mirror systems collecting sunshine to fire a boiler require enormous space and amazing precision; photocells which convert light directly to electricity are said to be economically feasible, but they don't resolve the problem of storing electricity. Other research areas covered include producing hydrogen directly from sunlight to fuel cars, using the sun to grow plants which can be burned as fuel or decomposed to make methane, and the proposed 50-square-mile power satellite. Provocative, but sometimes slow-moving, this section is definitely limited to more advanced senior high social studies and science students. Part II, *Wind and Water Energy,* discusses the possibility of using energy from the wind. It notes the thin energy of wind and the problems windmills have presented: durable sails, strong enough to provide a continuous energy flow; and the huge number of mills which would be needed. The energy potential of water in two forms in also considered: energy from ocean waves and from ocean thermal energy conversion (OTEC). Efforts to make electricity generated from OTEC competitive with other fuels are described. Low-key and thorough; its main audience will be serious science students.

329. Solar Energy: Putting Sunshine in Your Life. 160 slides w/2 cassettes and 2 discs, 34 min., color, w/tchr's. guide. Science and Mankind, 1978. $139.50. Gr. 10–12.

Part I describes the project of a group of high school students in Colorado who planned and built a solar-heated greenhouse designed to grow food for their school. The slides show what decisions needed to be made (e.g., passive vs. active collection systems), how the division of labor worked, and the results. The technical quality of the slides is excellent, but they do not always show clearly what the narrator is describing. Since several diagrams at the beginning of the series are not labelled, the viewer is immediately faced with such terms as "thermal mass" without any explanations. A diagram showing the principle of an active solar air-heating system is finally shown halfway through Part I. The idea of cooperative learning about a solar application is well conveyed and the comments of the students are believable. However, this part moves quickly and unless viewers are already familiar with solar energy and its terminology, they will need considerable background before it is shown. Part II concentrates on the ways people have heated their homes with the sun, beginning with the Indians of Mesa Verde. It moves more deliberately, showing the basic elements of a passive solar heating system, the materials used to effect the heating, and examples of homes heated passively. The elements of an active solar system using both water and air collectors are shown, and an architect who combined both systems discusses how he did it. People who have lived with solar energy describe their own experiences, noting, for example, that solar energy can be competitive with electricity or propane if a major element such as the collecting panel is reasonably priced. Fine slides show a variety of solar homes and provide a better explanation of the principles of solar heating than the first part. A two-part sound filmstrip set, *Solar Energy*, is said to cover approximately the same ground and is available with cassettes or records from Crystal Productions, Airport Business Center, Box 11480, Aspen, CO 81611, for $39 or $21 per part. It was not available for preview at this time.

330. The Solar Generation. The Science of Energy Series. 16mm, 21 min., color, w/tchr's. guide. Stuart Finley, 1976. $350; $35/1 day, $52.50/1 week rental. Gr. 10–12.

The first in a ten-film series on the science of energy, this solar film covers new territory in rapid fashion, and while it does not take time to explain unfamiliar terms, some are defined in the accompanying study guide. It opens with a visit to a "solar generation" art class in a Maryland school which is heated and cooled by solar energy and where solar energy is part of the curriculum. Other examples of solar utilization are then shown: houses using various solar heating systems, an energy-saving government office building, an industrial complex using experimental solar heating. Some time is spent explaining the problems of solar-cell research for direct conversion of energy into electricity by way of visits to private and government labs. The first house to use solar energy for both heating and electricity is shown, as is the inevitable solar furnace. The film exhorts Americans to stop treating solar power as a toy and bring it to its full practical potential in the American energy picture.

331. Solar Power: The Giver of Life. Energy Sources: A New Beginning Series. ¾" videocassette or 16mm, 29 min., color, w/tchr's. guide. Univ.

of Colorado, Educational Media Center, 1975. $230 videocassette; $333 16mm; $2.50 tchr's. guide; $10/3 days 16mm rental; $10/3 days, $20/15 days, $30/30 days videocassette rental. Gr. 10–12.

Man is returning to an ancient goal: to catch the sun. Because fossil fuels are depleted, costly, and environmentally expensive, man is examining the possibilities of solar power. This film explains what solar energy is and its advantages (it is abundant, inexhaustible, nonpolluting, suited to small dispersed areas, and free) and disadvantages (primarily its diffuseness or low intensity and variability). Better than any other film on the topic, it explains that solar's usefulness depends on the cost involved in converting it to other useful forms and goes on to describe the four main conversion methods. Low temperature conversion is clarified by means of good diagrams and an interview with Dr. Dan Ward of the Solar Energy Application Laboratory at Colorado State University, whose aim is for simple, replicable, reliable solar systems. Three types of experimental solar homes are shown. High temperature thermal conversion for electric power generation is discussed. Photo-voltaic conversion involving the conversion of light energy into electrical energy which produces voltage is explained, and the reasons for its prohibitive cost are given. Chemical conversion by photosynthesis is briefly discussed. The film concludes that the basic research on solar energy has been done, but the federal government has not granted research funds or depletion allowances for solar power as it has for other energy sources. It stresses that the advantages of free operating cost should be remembered when criticisms of the initial high cost of solar installations are levelled. Mainly done in studio shots, lecture style, the film also has some outside footage.

332. Solar Slide Set. 24 slides, color, w/descriptions. Zomeworks Corp., 1975, 1976. $15. Gr. 10–12.

This inexpensive set shows examples of six solar heating systems and devices. The first five slides which show the beadwall passive heating system for a greenhouse would have been less perplexing if some diagrams of this system had preceded them. The remaining slides, however, provide a good look at the drumwall, skylids, water heaters, air loop storage, and the nightwall systems. It should be useful with senior high students in building trades or architectural drawing classes. Teachers will need to introduce the basic principles of solar space and water heating through diagrams or a text and use this set to illustrate working systems.

Water Power

NONPRINT

333. Building a Hydro-Electric Power Complex—Churchill Falls. 10 slides, color, w/written narration. Donars Productions, n.d. $7. Gr. 7–12.

This slide set, accompanied by a written narrative for each slide, provides the kind of experience gained from touring a building site—somewhat technical and awesome with an amazing array of facts. The text accompanying these slides is printed in both English and French; requests to put this narrative on tape must be submitted to the producer for approval. The set is recommended because of its price and because its subject, the Churchill Falls plant, a generating facility in central Labrador, is said to be the largest single-site producer of electricity in the western world. Students will need some introduction to the design of hydroelectric plants prior to viewing.

334. Waterground. 16mm, 16 min., color, w/tchr's. guide. Appalshop Films, 1977. $200; $20 rental. Gr. 7–12.

The power of water to do work is shown in this sympathetic and scenic documentary of a small grist mill owned and operated by Walter Winebarger in Meat Camp, North Carolina. Close-ups of water pouring onto a wheel explain better than words could how water provides energy. Winebarger describes in a regional drawl his satisfaction with a form of power others have abandoned and his reasons for staying in the business of grinding flour by water power. ("I don't put nothing in nor take nothing out.") Old photographs are used to tell the history of the mill, while a General Mills flour-mill manager explains the economic advantages of a large mill electrified by TVA. The falling water and Winebarger himself offer an engrossing look at a disappearing part of America's energy scene.

Wind Power

PRINT

335. Cosner, Shaaron. **American Windmills: Harnessers of Energy.** New York: David McKay, 1977. 50pp. $6.95. Gr. 4–8.

This is a very brief, illustrated history of how windmills have been used in America over the years, first to grind grain, then to pump water for Western settlers. Many windmill designs are shown, including those for the ill-fated windmill at Grandpa's Knob, Vermont, and designs for experimental windmills which are currently being tested for their potential as an energy source. The book's prime use will be in social studies classes since neither the drawings nor the text provides enough information for model building.

336. Brown, Joseph E., and Brown, Anne Ensign. **Harness the Wind: The Story of Windmills.** New York: Dodd, 1977. 109pp., $5.95. Gr. 7–9.

With many illustrations of the mills described, this book covers the history of windmills throughout the world, how they were designed, what made them run, what work they performed, and how they operate. The importance of windmills to the economy of Holland in pre-steam engine days and the part wind power played in the American West is described. The authors discuss the relatively new role of wind as a generator of electric power and the possibilities of wind power competing with or supplementing fossil and nuclear energy sources. The amount of historical detail and the number of new and unfamiliar terms makes this more suitable for junior high readers than most upper elementary students.

337. Kagin, Solomon S. **Buyers Guide to Wind Power.** Santa Rosa, Calif.: Real Gas and Electric Co., 1977. 14pp. $3.98 pap. Gr. 9–12.

Directed at anyone considering a wind energy conversion system (WECS) for personal use, this booklet explains how to analyze the site potential for a WECS, how to determine available wind, how to make an energy use profile, and the components of a wind system. It provides brief, clear information about wind power and its practical applications, showing the integral parts of a wind-conversion system. The publisher sells wind-conversion equipment and does both site assessments and installations of these systems.

338. Clews, Henry. **Electric Power from the Wind.** Norwich, Vt.: Enertech, 1974. 40pp. $2. Gr. 10–12.

Subtitled "a practical guide to wind-generated power systems for individual applications," this book provides an easy-to-follow account of how a modern wind-powered generating system works. The author, whose Maine home is supplied with energy generated by two wind plants, explains the expected electrical output of a wind generator and provides tables to assist in making decisions about wind power. He discusses the power storage system he installed as well as the pros and cons of wind power. Sources of wind equipment are listed. This could be useful to a science or industrial arts class considering an experimental wind-power project.

339. Torrey, Volta. **Wind-Catchers: American Windmills of Yesterday and Tomorrow.** Brattleboro, Vt.: Stephen Greene Pr., 1976. 226pp. $12.95. Gr. 10–12.

Bringing together windmill lore and anecdotes from science, history, and literature, Torrey, a former *Popular Science* editor and wind machine buff, misses few items of windmill history and development in England, Holland, and America. The last 50 pages cover recent research on wind energy conversion systems in the U.S. and abroad, noting that the greatest obstacle to the use of wind power may be the cost of storing electricity. Torrey suggests that fuel cells or the redox cell may provide the solution to this problem. For the student with a strong interest in inventions or Americana, preferably both.

340. Hackleman, Michael A. **The Homebuilt, Wind-Generated Electricity Handbook.** Mariposa, Calif.: Earthmind/Peace Pr., 1975. 194pp. $7.95 pap. Gr. 11-12.

The style of the handbook is informal, even colloquial ("Extra precautions are in order here becuz there is little room for the small mistakes"), but the content is geared to the technically oriented student with an advanced knowledge of electricity and a strong interest in experimenting with wind power. The photographs and diagrams included are instructive, but some will be put off by Hackleman's breezy style and cartoons.

341. Hackleman, Michael A. **Wind and Windspinners: A "Nuts 'n Bolts" Approach to Wind-electric Systems.** Mariposa, Calif.: Earthmind/Peace Pr., 1974. 138pp. $7.95 pap. Gr. 11-12.

Hackleman spells out what is involved in making a wind-electric system. He notes the principles of wind energy; how to generate electricity; batteries; control systems; and how to use electricity with the S-Rotor aeroturbine, an alternate to propeller-type wind machines. Vocational education students in aircraft, carpentry, electricity, and electronics classes could use this as a resource for wind-generating projects.

342. Water and Wind Engines. 1 silent filmstrip, 32 fr., color. Standard Projector and Equipment Co., 1966. $7. Gr. 6-9.

This strip traces the use of the wind for sailing and exploring, but it does not advocate wind as a viable energy source. There is a good schematic explanation of a windmill's operation. The three basic kinds of water wheels are explained well, as are the relationships between water flow and energy. These explanations lead into diagrams and a discussion of how a hydroelectric plant operates and the importance of hydroelectric power. This format could be helpful with slower classes who might benefit more from reading the explanations at their own pace than from having the text read to them.

343. Wise Masters of Wind and Water. videocassette or 16mm, 16½min., color. Perspective Films, 1976. $238 videocassette; $264 16mm; $26 rental. Gr. 7-12.

This nonnarrated film, set in contemporary Hungary, is a visual tribute to the power of wind and water. It conveys to viewers the many ways in which people have traditionally used the simple power of water and wind to perform daily tasks. Showing a windmill grinding grain, and water wheels turned by a stream running a lathe, sharpening tools, turning food on a grill, sawing planks, and spinning yarn, the film demonstrates the importance of the stream—nature's power line. The only sounds are the creaking and pounding of the mill and its millstone and the sound of running water doing work. This could be used with general science classes to explain what it would be like to live in a low-energy world dependent on only wind and water power.

344. Wind Power: The Great Revival. Energy Sources: A New Beginning Series. ¾" videocassette or 16mm, 29 min., color, w/tchr's. guide. University of Colorado, Educational Media Center, 1975. $230 videocassette; $333 16mm; $2.50 tchr's. guide; $10/3 days 16mm rental; $10/3 days, $20/ 15 days, $30/30 days videocassette rental. Gr. 9–12.

Against the background of ranching water problems, the windmill became the prime means of supplying water to the Great Plains. Used in the past to grind corn, crush ore, make gunpowder, and pump oil, wind power is now being thought of as a way to generate electricity. This studio-based film uses rear projection and diagrams combined with some footage at an experimental Colorado wind-power project to examine the potential of wind power today. Wind's advantages—it is clean and nondepletable—and its shortcomings—such as variability and lack of information about generating costs—are mentioned. Studies by the National Aeronautics and Space Administration and the National Science Foundation and proposals for a forest of wind generators for the Great Plains and floating wind-power stations in the Atlantic are noted, as are the conditions necessary to make effective use of this resource. The lecturer asks why research on this unsophisticated yet workable resource has been shortchanged in favor of much more complicated technologies.

345. Wind Energy. ¾" videocassette or 1" or ½" videotape, 27:45 min., color. Nebraska Educational Television Council for Higher Education, 1976. $225; $35/7 days rental. Gr. 10–12.

Dr. William Hughes, head of the School of Electrical Engineering, Oklahoma State University, discusses possibilities of exploiting the wind's energy and notes that wind energy is not a simplistic resource. He notes the random quality of wind, suggesting it be used in complement with solar energy since they vary inversely in intensity. He reviews three major wind experiments currently under way and refutes the idea that wind energy is free. He believes that in some areas wind power may already be competitive with fuel in cost. A studio-based educational-TV production, the film uses footage of the techniques and systems described as well as many diagrams to give extremely thorough coverage of wind power, its possibilities, idiosyncracies, and problems.

Wood

PRINT

346. Wade, Harlan, and Wrigley, Denis. **Wood.** Milwaukee, Wisc.: Raintree Childrens Books, 1977. 32pp. $4.49 plb. Gr. K-2.

This clean and simple brown-and-yellow picture book explains in very few words where wood comes from, what it can be made into, how it burns, and why we must use it wisely. It could be used to introduce the idea of wood as an energy source.

347. Sherman, Steve. **The Wood Stove and Fireplace Book.** Bryn Mawr, Penn.: Bryn Mawr Pr., 1976. Dist.: Stackpole Books. 128pp. $5.95 pap. Gr. 7-12.

Noting that until the mid-19th Century up to 90% of the energy expended in the United States came from wood, the author makes a good case for growing, cutting, and burning wood for heating and cooking. Chapter three, which compares wood with coal, gas, and oil, and chapter nine, which explains how to get the most out of wood fuel by careful placement of the stove or fireplace and the utilization of other energy-saving techniques, are particularly helpful. Fire hazards are noted; specific stoves and fireplaces are described and priced. The illustrations do not provide optimum instruction.

348. Eckholm, Erik P. **Losing Ground: Environmental Stress and World Food Prospects.** New York: Norton, 1976. 223pp. $7.95; $3.95 pap. Gr. 9-12.

Chapter six, "The Other Energy Crisis: Firewood," alerts us to the widespread and growing shortage of firewood for cooking purposes in Asian and African countries and its tragic results for people who have no alternate energy source. Eckholm traces the problem to the phenomenal growth in population over the last decade and to the quintupling of oil prices which took kerosene out of the reach of the world's poor. As a result, forest lands are stripped and the use of dung for fuel has become common, robbing the soil of nutrients and organic matter and leading to erosion and flooding. Eckholm's approach to solving these problems is realistic. This chapter was published separately by the Worldwatch Institute, where Eckholm is a senior researcher (see No. 349).

349. Eckholm, Erik. **The Other Energy Crisis: Firewood.** Worldwatch Paper No. 1. Washington, D.C.: Worldwatch Institute, 1975. 22pp. $2. Gr. 9-12.

This paper appeared in almost the same form as chapter six of Eckholm's book, *Losing Ground: Environmental Stress and World Food Prospects* (see No. 348).

350. United Nations Environment Programme. **Report of the Executive Director of UNEP on the State of the Environment, 1977.** New York: United Nations Environment Programme, 1977. 33pp. free. Gr. 9–12.

Section five of this report, "Use and Management of Renewable Resources: Firewood," notes the importance of firewood as an energy resource in developing countries and explains the need for growing firewood on a sustainable basis. One table compares firewood and charcoal consumption in various parts of the world.

351. Gay, Larry. **The Complete Book of Heating with Wood.** Illus. by Russell Stockman. Charlotte, Vt.: Garden Way, 1974. 128pp. $3.95 pap. Gr. 10–12.

Larry Gay, a scientist and advocate of heating with wood, brings good news: the total amount of wood in the U.S. is increasing every year, wood is low on the list of environmental polluters, and wood *can* be cheaper than gas, oil, or electricity for heating. He writes in a literate, knowledgeable style about the choice of woods, cutting and hauling wood, and utilizing wood properly. ("To burn wood is easy, but to burn it efficiently is another matter.") Ben Franklin is quoted frequently in the section on stoves, which is well detailed. A final section discusses proper ventilation and the location of wood burners, energy waste, and a variety of other topics.

352. Self, Charles. **Wood Heating Handbook.** Blue Ridge Summit, Penn.: Tab Books, 1977. 235pp. $5.95. Gr. 10–12.

Noting that the average masonry fireplace without accessories to increase heat is no more than 10-20% efficient while a wood stove does considerably better, the author estimates that homeowners can save up to 100% of their heating bill by using wood. He explains how to select, use, install, and maintain fireplaces and wood stoves and provides detailed advice on finding, choosing, cutting, splitting, seasoning, and burning wood. With its good, clear illustrations, this book could be used in units on wood heating, in refrigeration, heating and air-conditioning courses, or by students who want to improve heating efficiency in their homes.

353. Vivian, John. **Wood Heat.** Illus. by Liz Buell. Emmaus, Penn.: Rodale Pr., 1976. 320pp. $8.95; $4.95 pap. Gr. 10–12.

Vivian states that as of 1976 wood, "the only truly renewable fuel source," is easily the most economical energy source for the one-third to one-half of American families living in rural, small town, and suburban areas near forested areas. With the aid of good, simple illustrations, he describes and explains chimneys, heating with wood stoves, building new fireplaces and improving old ones, and cooking with wood heat. Though this is basically a how-to book, chapter one, "The Science and History of Wood Heat," is useful as a brief history of fuel use in America. A chapter called "Getting the Wood In" includes detailed instructions on buying and cutting wood, and a section on woodlot management could be used in conservation or horticulture classes. A list of sources for stoves, fireplaces, and central heating units is appended.

354. Shelton, Jay and Shapiro, Andrew B. **The Woodburners Encyclopedia; An Information Source on Theory, Practice and Equipment Relating to Wood as Energy.** Illus. by Vance Smith. Waitsfield, Vt.: Vermont Crossroads Pr., 1976. 155pp. $6.95 pap. Gr. 11–12.

Though this work in not arranged like the typical encyclopedia, it is encyclopedic in the thoroughness of its coverage of wood as an energy resource. Shelton explores the potential of wood as a direct energy source in rural and wooded areas or for power plants, as a source convertible to liquid or gaseous fuels for use in existing furnaces and engines, and as a source of industrial chemicals. In 13 technical chapters, he discusses the concepts of energy, temperature, and heat; the energy content of wood; combustion; chimneys; the energy efficiency and operating characteristics of wood stoves; and the effect of installations on the usefulness of wood stoves. There are brief chapters on stove accessories, fireplaces, and chimney fires. A closing chapter compares the cost of woodburning for heat with the costs of oil, natural gas, electricity, and liquefied petroleum. Section two contains a detailed list of manufacturers and importers of stoves, fireplaces, and accessories, with descriptions supplied by the vendors. Section three contains manufacturers' specification charts.

Part Two

Curriculum Materials

Curriculum Materials

355. Bakke, Ruth. **Energy Conservation Activity Packets.** 5 packets. Des Moines, Iowa: Iowa Energy Policy Council, 1977. approx. 60pp. ea. $10/set of 6; $2 ea; free to Iowa teachers after in-service training. Gr. K–2; 3; 4; 5; 6.

Designed to aid teachers in encouraging the development of a conservation ethic by their students, this program utilizes a values clarification approach. Each of the five packets is in the form of a plastic-covered notebook and contains a variety of activities with the goal of producing both energy learning and an awareness concept. In the primary grade packet, emphasis is also given to involving parents in the learning process. Annotated children's and teacher's bibliographies are included in each packet.

356. Virginia Energy Office. **Energy in the Classroom.** 3 vols. Richmond, Va.: Virginia Energy Office, 1975. $8/set; Vol. 1, 38pp., $2, Gr. K–3; Vol. 2, 102pp., $3, Gr. 4–7; Vol. 3, 143pp., $3, Gr. 8–12.

Three activity guides make up this series. Volume one includes things to do and talk about, experiments and demonstrations, puzzles, resources for students and teachers, and simple definitions. Volume two introduces the energy crisis and our major energy sources (followed by suggestions for class discussion and activities) and suggests energy demonstrations, conservation activities, and creative energy activities. A 25-page teacher's resource section is included. Volume three suggests classroom activities and discussion areas for 19 different subject areas ranging from Sociology to Driver Education. A 50-page teacher's resource section is included.

357. Jamason, Barry W., ed. **Living within Our Means: Energy and Scarcity.** Albany, N.Y.: New York State Education Dept., 1974. 2 vols. Vol. 1, 86pp. Vol. 2, 106pp. free to New York State teachers; available to others through ERIC. Vol. 1, Gr. K–6; Vol. 2, Gr. 7–12.

In both of these instructional manuals, energy-related activities are suggested by grade levels within the mathematics, language arts, science, and social studies disciplines. According to the editor, each book is meant to illustrate the most appropriate and efficacious manner in which environmental concerns may be integrated into classroom instruction. And so, environmental education activities are grouped in accordance with their relationship to the subject matter of the various disciplines. These relationships are reinforced by references to the pertinent skill, objective, or understanding in the New York State subject matter curriculum bulletins. In this way, the syllabus references should justify the placement of the environmental activities within the regular instructional framework.

358. Coon, Herbert L., and Alexander, Michele Y. **Energy Activities for the Classroom.** Columbus, Ohio: ERIC/SMEAC Information Reference Center, 1976. 148pp. $4.50 (Item 016E). Gr. K–12.

This sourcebook, designed for use in grades K–12, contains energy teaching activities related to energy resources, production, distribution, and use. Each activity has been classified by the editors for the most appropriate grade level, subject area, and energy concept involved. Subject areas are science, mathematics, social studies, language arts, and fine arts. References cited in specific activities are useful to persons interested in obtaining more information on activities and ideas related to energy. Many of the activities are interdisciplinary in nature and were developed or suggested by public school teachers. The book is arranged K-3, 4-6, 7-9, 10-12, with about 100 pages pertaining to grades 7 and up.

359. Energy Crisis: A Teacher's Resource Guide. Trenton, N.J.: New Jersey Education Assn., 1974. 12pp. free. Gr. K-12.

Prepared through the combined efforts of the New Jersey State Department of Education, New Jersey State Council for Environmental Education, and New Jersey Education Association, this pamphlet presents a series of 11 objectives, background content, activities, references, and examples of measuring devices pertaining to the energy crisis.

360. Mengel, Wayne. **Energy, Key to the Future: Teaching Techniques for the Understanding and Conservation of Energy.** Poughkeepsie, N.Y.: Dutchess County Board of Cooperative Educational Services, 1974. 32pp. $4. Gr. K-12.

The teaching techniques presented in this booklet are designed to provide students with concepts which relate to the energy crisis and to energy conservation. The techniques are *not* presented in the form of completed lesson plans. Rather, they are ideas that act as starting points for further development by the teacher. It is expected that the teacher will select and develop those techniques which best meet the needs and ability of the student. Twenty-two energy-related projects are listed, as are energy conservation tips and a glossary.

361. Michigan Association of School Administrators, Region 9. **Energy Conservation: Guidelines for Action.** Lansing, Mich.: Michigan Assn. of School Administrators, 1974. 58pp. $.75. Gr. K-12.

Subtitled "Suggested guidelines for local school district development of energy conservation programs," this publication contains background information on conservation, suggested guidelines for the study of conservation, representative activities and resources, and guiding principles for establishing programs. The appendix contains an overview of energy resources.

362. New Jersey Department of Education. **Energy and Education Handbook.** Eugene, Ore.: ERIC Clearinghouse on Educational Management, 1977. $10.03; $.83 microfiche (ED 144-261). Gr. K-12.

Presented at the Energy and Education Conference for New Jersey school personnel in October, 1977, this resource book discusses the need for energy education and conservation, as well as energy considerations in relation to the physical facilities of schools, and it includes curriculum models, a bibliography, and a glossary. The emphasis is on conservation of energy.

363. Meaney, Marie. **Home Sweet Earth: An Environmental Learning Experience for First Grade Level.** Seattle, Wash.: Highline Public Schools, Project Ecology, 1973. unp. $2.50. Gr. 1.

This ecology unit is designed to help the first grade child become aware of the many wonders of nature and to learn something about the energy and ecology problems of the present and future. Each of ten lessons contains a concept, materials list, procedure, evaluative activity, and suggested additional activities. Lessons one, two, three, and nine are directly related to energy.

364. National Science Teachers Association. **Interdisciplinary Student/ Teacher Materials in Energy, the Environment, and the Economy.** 6 units. Oak Ridge, Tenn.: U.S. Dept. of Energy, Technical Information Center, 1977. pp. vary. free (send requests for bulk orders on school letterhead). Gr. 1; 2; 8, 9; 8, 9; 9, 11, 12; 10, 11, 12.

Produced by the National Science Teachers Association Project for an Energy-Enriched Curriculum, this series includes six instructional units at various grade levels. Descriptions of the units follow.
1. *The Energy We Use* (Gr. 1) is a set of nine lessons which relate closely to the traditional curriculum for this grade level—the student and his/her world. Each lesson provides an objective, a list of materials, background information for the teacher, teaching strategies, and a summary. The unit encompasses energy from food, the sun, fossil fuels, wind, and water, and the positive and negative aspects of these sources.
2. *Community Workers and The Energy They Use* (Gr. 2). Activities are intended to blend a child's knowledge of energy with the study of the effect energy use has on the people in the community. The unit comprises 13 lessons divided into four categories: introduction to energy; community workers who work directly with the sources of energy; community workers whose work depends on a continual supply of energy; and community workers who make decisions about energy.
3. *Transportation and the City* (Gr. 8, 9). The learning activities of this packet are designed to fit into existing segments of instruction in U.S. history and civics. Four lessons introduce students to automobiles, their operational costs, and their relationship with mass transit; study the effects of the car on the American small town; present information about the car-centered city of Los Angeles; and put the automobile on trial.
4. *Energy, Engines, and the Industrial Revolution* (Gr. 8, 9) examines the broad social and economic upheavals of the Industrial Revolution and demonstrates the close link between machinery and the abundant use of energy. There are five lessons in all. In two science lessons, students describe the energy conversions which take place in an automobile and consider the engine upon which the Industrial Revolution was built. The remaining three lessons are in the social studies. "Student readers" which support each lesson make up half the unit's content.
5. *How a Bill Becomes a Law to Conserve Energy* (Gr. 9, 11, 12). The goal of these materials is to integrate facts and concepts about energy, environment, and economics into the study of the process of making and applying

a law, in this case, the 55-mile per hour speed limit. Six activities are included.

6. *Agriculture, Energy, and Society* (Gr. 10, 11, 12) proposes to help students to examine present-day agricultural methods and to study the impact of these methods on existing energy resources. The eight activities are designed to fit into existing segments of instruction in social studies, mathematics, and science taught at the secondary level. One activity is a study of energy efficiency in corn production; another uses graphs and tables to show how American farming has changed from labor-intensive to energy-intensive.

365. Energy and Man's Environment. **Energy and Conservation Education: Activities for the Classroom.** 4 notebooks. Portland, Ore.: Energy and Man's Environment, 1977. pp. vary. $86.40/set; $24 ea. Gr. 1–3; 4–6; 7–9; 10–12.

This series of learning-activity packages was designed to present practical and objective energy and conservation education classroom resources. There is a loose-leaf notebook for each grade span. Each notebook is divided into six energy-related sections: sources, uses, conversion, impacts, limits, and the future; and each section has its own learning goal. The concept, objective, timing, subject area, implementation, materials needed, and suggested assessment procedure are provided for each activity within a section. The assessment section in each notebook includes test items which are multiple choice, true/false, completion/fill-in, or discussion questions, depending on the grade level. Answers are provided for teachers.

366. National Science Teachers Association. **Award Winning Energy Education Activities.** Washington, D.C.: National Science Teachers Assn., 1977. 38pp. free. Gr. 2–12.

This booklet contains brief descriptions of the winning entries in the NSTA Teacher Participation Contest conducted in spring, 1976. This contest was part of a project in energy education sponsored by the Education Programs Branch of the Energy Research and Development Administration; its goal was to locate ideas for activities which would fit easily into standard courses of study and at the same time further student understanding of important energy issues. The group includes an aluminum-recycling experiment for a senior high chemistry class, a role-playing exercise to acquaint students in grades 7–12 social studies or science classes with the problems of securing new oil supplies, and "Kill a Watt," a series of eighth grade math activities to encourage energy conservation.

367. Bonneau, Marilyn, and Lee County Environmental Education Program. **Energy: Where Do We Get It and How Do We Use It?** Student Learning Activity Module. Fort Myers, Fla.: Lee County Schools Environmental Education Program, 1974. unp. $.25. Gr. 3.

A short two- to five-day teaching unit that can be used to engage students in environmental education. This module was developed to help young students become more aware of energy: how it is produced, how we use it, and some forms of energy that may be used in the future. It contains lesson plans, a coloring booklet, pre- and post-tests, and activity and discussion suggestions. Four lessons are included.

368. Project CLEAN Curriculum Modules. Shawnee Mission, Kan.: Shawnee Mission Public Schools, n.d. 4 modules. 14–46pp. ea. $1.50 ea. Gr. 4; 7; 9–12; 7–12.

Four of the modules from this Title III project (*Cooperative Learning through Environmental Activities in Nature*) are related to the study of energy. Each module contains the following components: description of the scope, objectives stated in behavioral/performance terms, a set of lesson plans, a content-valid test encompassing all objectives, suggestions for use and implementation of the module, a list of relevant audiovisual materials, and references to outside resources which can be used in implementing the module. The relevant modules are:
1. *The Historical Development of Water Power,* developed by Sharon Spector. Grade 4 or intermediate students who are mentally retarded or have learning problems. This is an exploration of the development and the resulting ecological implications of water power. Also explored is water power's role in today's industrial complex.
2. *The Energy Problem,* developed by Hal Jehle. For use in an environmental unit of grade 7. This unit aims to acquaint students with the depletion of our fossil-fuel resources, its possible consequences and future alternatives.
3. *Crisis in Power,* developed by Karen Smith. Grades 9 through 12. Written for use in science and social studies courses, the module proposes to allow students to examine the energy needs of the United States and the world and to consider possible energy resources, evaluating the impact of each on the environment.
4. *Changes Ahead for a Gas-Guzzling America,* developed by Jim Tilley. Grades 7 through 12. This module attempts to make students aware of changes that must occur in their lives because of gasoline shortages and higher prices.

369. Ward, Barbara T., and Lee County Environmental Education Program. **Conservation of Energy.** Fort Myers, Fla.: Lee County Schools Environmental Educational Program, 1974. 13pp. $.40. Gr. 4.

The purpose of this unit is to teach children the origin and development of electricity, to make them aware that our sources of power are endangered, and to help them to realize that they can have a part in protecting the environment through their words and acts. Included are five lessons, pre- and post-tests, individual and class activities, and pages from which to make transparencies, with an explanation of each page.

370. Oak Ridge Associated Universities. **Science Activities in Energy.** Oak Ridge, Tenn.: The American Museum of Atomic Energy/Oak Ridge Associated Universities, 1977. unp. free. Gr. 4–6.

This is a series of simple, concrete experiments developed to illustrate certain principles and problems related to various forms of energy and their development, use, and conservation. The four items in the series are: Chemical Energy, Electrical Energy, Solar Energy, and Conservation. Each of these is a two-color folder containing between 12 and 16 heavily illustrated experiments which present a coherent program on the topic. Metric measurements are used, and with few exceptions, easily available materials are suggested. Each experiment is contained on a separate page for duplication or projection. Notes to the teacher explain the purpose of each experiment.

371. Jones, John, ed. **Energy and Man's Environment: Activity Guide.** 7 booklets. 2nd ed. Portland, Ore.: Energy and Man's Environment, 1976. 33-73pp. ea. $25/set. Gr. 5-9.

The energy *Activity Guide* consists of seven separate booklets: an introduction, sources of energy, uses of energy, conversion of energy, impacts of energy, limits of energy, and future sources of energy. The set is interdisciplinary and is designed for classroom teachers and curriculum writers involved in preparing learning experiences relating to energy and the environment. Each booklet contains concepts which the producer feels are necessary to a fundamental understanding of the topic; each concept, in turn, has a number of objectives and activities to insure learning. In toto, the set contains several hundred energy- and environment-related instructional ideas and activities.

372. Gandy, Sharon S., and Lee County Environmental Education Program. **The Energy Crisis: Social Studies or Interdisciplinary.** Fort Myers, Fla.: Lee County Schools Environmental Education Program, 1974. 19pp. $.40. Gr. 6.

This module was developed to create an awareness of how technology can improve life but also create problems. On the completion of this unit students should have an understanding of the effects of intelligent versus wasteful use of natural resources, and how modification of the natural environment may have both good and undesirable consequences. Very little special preparation is needed for the module, which can be team taught by all four disciplines, taught in conjunction with language arts, or taught only as a social studies unit. This five-lesson unit contains brief lesson plans, a pre-/ post-test, audiovisual aid suggestions, and a student module with six activities, projects, and extra-credit assignments.

373. Hughes, Judi, and Lee County Environmental Education Program. **The Energy Crisis: The Dictionary as Resource.** Fort Myers, Fla.: Lee County Schools Environmental Education Program, 1974. 22pp. $.45. Gr. 6.

This five-lesson module was developed to help create an awareness of the causes of and some solutions for the energy crisis. The unit utilizes the dictionary and dictionary skills as a mode for working with the energy crisis problem, and it could be used as an introduction to dictionary skills. Lesson plans, resource materials, and a teaching module are included.

374. Sheridan, Jack. **Investigating Resource Acquisition and Use.** Houston, Tex.: Harris County Dept. of Education, 1977. 108pp. $2.50. Gr. 6-9.

This teacher's notebook is one of 11 activity-based units for teaching environmental education through science and social studies; the lessons are designed to help students acquire environmental concepts and values through activities involving indoor and outdoor investigations and simulation exercises. The goal of this particular unit is to help students achieve insight into decision-making in resource acquisition and use. Petroleum was selected for this experience because the problems associated with its acquisition and use were thought similar to those of other resources which may be less familiar to the student.

375. Hennings, George, and Hennings, Dorothy Grant. **Keep Earth Clean Blue and Green: Environmental Activities for Young People.** New York: Citation Pr., 1976. 250pp. $9.95; $5.95 pap. Gr. 6-10.

Chapter seven, "Energy Challenges: Activities with Energy Problems," offers ideas for learning activities about coal; nuclear, hydroelectric, geothermal, and solar power; and power from the tides, from plants, and from garbage. Other activities center on energy consumption—both worldwide and local— and conservation. Sources are suggested. Activities vary in complexity. Some are appropriate for science, others for social studies.

376. Parr, Donald. **Energy Futures: An Environmental Learning Experience for Junior High Science.** Seattle, Wash.: Highline Public Schools, Project Ecology, 1973. unp. $2.50. Gr. 7-9.

This "Environmental Learning Experience" was prepared with the following assumptions: "Based on today's consumption rates and known reserves, the seventh grade student will be 34 or 35 years old when we run out of oil products. He or she will be 22 years old when we run out of natural gas. The student has three possible futures: 1) he or she may live in an altered lifestyle designed to prolong our energy sources by conserving energy; 2) he or she may maintain the present lifestyle if renewable energy sources are developed; or 3) he or she may live a completely different lifestyle if we run out of usable energy. The choice is not entirely up to the student but he and she will have the responsibility of implementing decisions made now and of making new decisions about energy use." Eleven lessons, each taking two to three weeks to complete, are developed. They cover topics such as how much energy the student and his/her family uses, renewable energy sources, and energy conversion in electricity generation. Each lesson is divided into concept, materials, procedure, evaluative activity, and suggested extra activities. A materials and equipment list is provided.

377. University of Tennessee Environment Center. **Ideas and Activities for Teaching Energy Conservation: Grades 7-12.** Knoxville: Tenn.: Univ. of Tennessee Environment Center, 1977. Dist.: Tennessee Energy Authority. 225pp. $2. Gr. 7-12.

This multidisciplinary curriculum guide provides background information on energy resources, consumption, and conservation, followed by more than

100 pages of activities, divided by discipline: science, social studies, communications/language arts, and multidisciplinary. Each discipline is divided into activities for grades 7-9 and 10-12. Appendices include sources of additional information, energy conversion tables, a glossary, and a bibliography.

378. Pennsylvania Department of Education, Bureau of Curriculum Services. **The Environmental Impact of Electrical Power Generation: Nuclear and Fossil.** Washington D.C.: U.S. Government Printing Office, 1977. 89pp. $3.25 (ERDA-69). teacher's manual, $1.25 (ERDA-70). Gr. 9-12.

A mini-course for secondary schools and adult education, the intention of this working paper is to provide as unbiased and straightforward a view of the advantages and disadvantages of nuclear generation of electrical power as possible. Written and compiled by an independent committee made up of educators, engineers, health physicists, members of industry and conservation groups, and environmental scientists, chapters describe the biological and environmental effects of nuclear and fossil-fueled power plants. An 18-page glossary of relevant terms is included.

379. The Solar Energy Project. Albany, N.Y: New York State Education Dept., Bureau of Educational Communications, n.d. price not available. Gr. 9-12.

This solar energy curriculum includes activities, articles, bibliographies, and "Topics on Tap"—lists of energy-related materials in the collection of the New York State Library, as compiled by the Legislative Service of the Library.

380. Terry, Mark, and Witt, Paul. **Energy and Order or If You Can't Trust the Law of Conservation of Energy, Who Can You Trust?** San Francisco, Calif.: Friends of the Earth, 1976. 42pp. $3. Gr. 9-12.

Written by two tenth grade teachers, this program supplies an energy vocabulary and seeks to develop the student's ability to analyze the energy and order relationships of a system and its surroundings. It emphasizes practical analyses of systems and issues important in our lives: the kitchen, the automobile, the supermarket, the workplace, economic inflation, population growth, and per capita energy growth. Planned as a four-five week unit in a biology course, this program is intended to be implemented with suggested audiovisual materials, a hands-on activity session with each section, and follow-up section discussions.

381. University of Tennessee Environment Center, and College of Home Economics. **Energy Conservation in the Home; An Energy Education/Conservation Curriculum Guide for Home Economics Teachers.** Oak Ridge, Tenn.: U.S. Dept. of Energy, Technical Information Center, 1977. 319pp. free (send requests for bulk orders on school letterhead). Gr. 9-12.

This guide was designed for home economics teachers as a source of information, instructional materials, and suggested references about the energy situation as a whole, energy concepts, and the use of energy in the home.

The guide attempts to provide a synthesis of current energy information from a multitude of sources which are not readily available to home economics teachers; it seeks to better prepare the teacher to instruct students, especially about their role in conserving energy. An overview of the energy situation is combined with a complete guide to energy uses and practices in the home. In loose-leaf form, with an expanded table of contents, the book's divisions correspond to the most commonly used curriculum divisions for home economics: foods, clothing, housing, personal care. Educational objectives are provided for every major division as well as each activity; twenty-six activities are provided. Key terms are italicized and defined within the text and in a glossary. An extensive bibliography is included.

382. Changing Times Education Service. **This "Energy Crisis": Is It Real?** Washington, D.C.: Changing Times Education Service, 1972. $12. Gr. 10-12.

A CTES inquiry-oriented mini-unit, one of 53 highlighting important consumer problems for analysis, this contains 40 copies of a four-page article of the same name from *Changing Times,* an overhead transparency and spirit master depicting in bar-graph form the U.S. energy forecast (total demand and expected sources of supply through 1985), and a guide to teaching this mini-unit. The teaching guide lists exercises for review, inquiry, and discussion.

383. Educational Research Council of America, Science Department. **What's the Matter with Energy? (Energy Sources).** 1st experimental ed. Cleveland, Ohio: Educational Research Council of America, 1974. teacher's guide, 39 pp., $2.05. student's guide, $1.84. Gr. 10-12.

This program makes an excellent high school level mini-course in response to requests for an investigation into the energy crisis. The class first views and discusses a film about energy. The remainder of the "Introductory Activity" is a resource section on measurement skills that students can refer to as they do the "Alternate Activities." Different energy sources and some of their interactions are individually explored in the nine "Alternate Activities." During the "Generalizing Activity," the class investigates different societies and their energy needs and discusses the current state of the energy crisis.

384. Jamason, Barry W. **Environmental Quality: A Community Concern.** Albany, N.Y.: New York State Education Dept., 1977. 50pp. free to N.Y. State teachers; not available to others. Gr. 10-12.

This collection of four seminar topics is suggested as a co-curricular offering, as the core of an environmental studies option (mini-units or a semester elective), or, using only particular units, it can be used within the context of instruction in one of the academic areas. Each topic comprises a seminar description, background information, suggested procedures or activities, and worksheets. The relevant topics are energy consumption, and to a lesser degree, transportation.

385. Norton, Thomas W. **Solar Energy Experiments for High School and College Students.** Emmaus, Penn.: Rodale Pr., 1977. 129pp. $5.95. Gr. 10–12.

This manual is suggested as a solar energy curriculum outline. It includes 19 self-contained experiments of varying difficulty which emphasize energy measurements and eight classroom activities which present worldwide energy data and solar energy data for individual student analysis. The author is a research associate at the State University of New York (Albany) Atmospheric Sciences Research Center, as well as a high school physics instructor.

386. Pennsylvania Department of Education, Bureau of General and Academic Education. **Nuclear Science: A High School Course.** Harrisburg, Pa.: Pennsylvania Dept. of Education, Office of Information and Publications, 1971. 119pp. $.85; free to Pennsylvania public school educators. Gr. 10–12.

Designed to provide a basic but comprehensive understanding of the atom, this course pays special attention to the use of the energy of the atom as a contributing factor to modern civilization. Crossing what are considered the normal disciplines of science, it seeks to develop student appreciation of the problems which confront our civilization, such as nuclear disarmament, the health problems of the atomic age, and survival in a nuclear war. The course also shows how man attempts to harness the tremendous energy liberated through fission and fusion reactions to provide the power necessary for the civilization of the future. Sections list sources of radioisotopes, textbooks, references, audiovisual aids, and sources of equipment. The complete course of study—eleven units with time span, overview, objectives, content, and suggested activities for each—is outlined.

387. U.S. Department of Energy. **Activities of the Department of Energy in Energy Education; A Description of Programs for Schools of the Department of Energy and Its Predecessor Agencies.** Washington: U.S. Dept. of Energy, 1978. Dist.: U.S. Dept. of Energy, Technical Information Center. 66pp. free. Professional.

The Department of Energy's responsibilities in implementing the National Energy Plan include identifying the appropriate role of educational activities and institutions in that work. This report describes five major areas in which DOE and its predecessor agencies have played a significant role: training, curriculum development, educational special events, facilities support, and the Energy Extension Service. Although research activities conducted at universities are excluded from the report, activities which have taken place at almost every level of education are included. Of particular interest is the section "Educational Special Events," which notes contests, fairs, competitions, conferences, workshops, and seminars sponsored by DOE; reports on a number of these events are available to administrators and curriculum planners. Sources are noted throughout.

Part Three

Selection Aids

Selection Aids

PERIODICALS

388. Appraisal: Children's Science Books. Children's Science Book Review Committee, Longfellow Hall, 13 Appian Way, Cambridge, MA 02138. 1967. 3/yr. $6.

This selection guide to children's and young adult science books is prepared by the Children's Science Book Review Committee, a nonprofit organization sponsored by the Harvard Graduate School of Education, the New England Round Table of Children's Librarians, and the New England Library Board. Bibliographic data and suggested age level are provided for each item, followed by two separate reviews prepared by a children's librarian and a science specialist. Ratings from excellent to unsatisfactory are also indicated. Reviewers are identified at the back of each issue.

389. Booklist. Paul Brawley, ed. American Library Association, 50 E Huron St, Chicago, IL 60611. 1905. 24/yr. $24.

Reviews adult and juvenile trade books and nonprint materials "worthy of consideration for purchase by small and medium sized public libraries, school media centers, and community college libraries." Titles selected by the staff as particularly noteworthy are starred. Adult books related to energy would be found under such headings as Social Sciences, Pure Sciences, and Technology. Young adult energy titles are listed under the subheading Nonfiction. Children's books are listed alphabetically by author and include grade level. There is also an Easy Reading section. Nonprint materials are arranged by medium: films, filmstrips, multimedia kits, slides, video, and recordings. Reviews are prepared by the *Booklist* editorial and reviewing staff and by selected librarians, media specialists, and subject specialists.

390. Bulletin of the Center for Children's Books. Zena Sutherland, ed. University of Chicago Graduate Library School, University of Chicago Press, 1100 E 57th St, Chicago, IL 60637. 1945. 11/yr. $10.

Provides between 700 and 800 analytical reviews of children's and young adult books, both fiction and nonfiction, each year. Reviews are arranged by author. The quality, usefulness, and potential audience of each title are carefully analyzed. All reviews are prepared by the editor, Zena Sutherland.

391. The Horn Book Magazine. Paul Heins, ed. Horn Book, Park Square Bldg, 31 St. James Ave, Boston, MA 02116. 6/yr. $12.

Reviews approximately 450 children's and young adult books and nonprint materials each year. A special section, "Views on Science Books," features reviews compiled by specialists in the field. General age levels are indicated for each title reviewed.

392. Instructor. Leanna Landsmann, ed. Instructor Publications, Instructor Park, Dansville, NY 14437. 9/yr. $14.

Each year *Instructor* publishes about 500 reviews of books, nonprint materials, and professional books suitable for use by elementary students and teachers. Science materials are reviewed in the October, December, January, and March issues; films and filmstrips in October, December, February, and April. A bibliography, "Free Energy Sources," appeared in the September, 1977 issue and an energy feature is scheduled for October, 1978.

393. Kirkus Reviews. Barbara Bader, ed. Kirkus Service, 200 Park Ave S, New York, NY 10003. 1933. 24/yr. Rates set by book budget.

Reviews adult and juvenile titles of general interest approximately two months before the books are published. A specialist in the field of science or economics and technology reviews adult titles related to energy. All children's books are reviewed by a children's books specialist; relevant energy titles will be listed under Younger Nonfiction or Older Nonfiction.

394. Landers Film Reviews. Bertha Landers, ed. Landers Associates, Box 69760, Los Angeles, CA 90069. 1956. 5/yr. $45.

A current guide to nontheatrical 16mm films thought to be worthy of consideration by public libraries, media centers, colleges, universities, and other users of nontheatrical films. Between 130 and 150 films are reviewed in each issue. Reviews, written by in-house professional staff, note the content and purpose of the film, film techniques used, and the intended grade level. Films are indexed by subject; relevant titles are listed under such subject headings as Conservation, Ecology, Energy, General Science, Physical Science, Social Problems, Industry, and Social Comment. Super 8mm films, filmstrips, slides, charts, graphic materials, and study prints are described; these items are arranged by distributor.

395. Library Journal. John N. Berry, III, ed. R. R. Bowker Co., Box 67, Whitinsville, MA 01588. 1876. 22/yr. $24.

Each issue of *LJ* features approximately 100 reviews of adult trade books prepared by outside librarians. The book review section is arranged by subject; energy titles suitable for use with students could be found under the headings The Contemporary Scene, Reference, Business and Economics, Political Science and International Affairs, Science and Technology, and Social Science. There is an author index to the book review section in each issue. For the past two years, the January 1st issue of *LJ* has included a special energy source directory listing organizations, government agencies, and nongovernment agencies, publishers, and companies who publish materials and/or conduct research on energy issues.

396. Previews. Phyllis Levy Mandell, ed. R. R. Bowker Co., Box 67, Whitinsville, MA 01588. 1972. 9/yr. $15.

Over 1300 nonprint materials for all age levels are reviewed in this journal each year. Reviews are prepared by audiovisual specialists, teachers, and librarians. Items are arranged by medium, e.g., 16mm films, filmstrips, slides, kits, games, etc., and then broken down by subject. Energy-related materials would be found under the subject headings Energy, Environment and Ecology, Social Studies, and Science. Grade levels are also indicated for each item.

397. School Library Journal. Lillian N. Gerhardt, ed. R. R. Bowker Co., Box 67, Whitinsville, MA 01588. 1954. 9/yr. $13.

In addition to a number of articles of interest to the library and teaching professions, *SLJ* features reviews of approximately 2500 children's, young adult, adult, and professional titles each year. Reviews of juvenile titles are prepared by librarians in the field. Appropriate energy titles can be found under the headings Preschool and Primary, Grades 3–6, and Junior High Up. A title and author/illustrator index is provided in each issue.

398. Science and Children. Phyllis R. Marcuccio, ed. National Science Teachers Association, 1742 Connecticut Ave, NW, Washington, DC 20009. 1963. 8/yr. $15; $17 to institutions.

Reviews approximately 500 books and audiovisual materials suitable for use in elementary school science classes. Curriculum materials are listed in the Curriculum Reviews section. Other books and audiovisual aids are listed separately in the Resources Reviews section. Reviews are prepared by 100 science educators across the country.

399. Science Books and Films. Shari Finch, ed. American Association for the Advancement of Science, 1776 Massachusetts Ave, NW, Washington, DC 20036. 1965. 4/yr. $16.

More than 1200 books and 16mm science films are reviewed annually in this publication. Of these, approximately 500 fall under the category Children's Books and Film Reviews. However, many titles in the main book review section are recommended for use with secondary school students. Reviews are arranged in Dewey decimal order. Reviewers include scientists or science specialists in high schools, colleges, universities, industry, libraries, and media centers. Each review is plainly coded to indicate the level of quality and the level of difficulty from kindergarten or preschool to professional. Title indices to books and films are included in each issue.

400. The Science Teacher. Rosemary Amidei, ed. National Science Teachers Association, 1742 Connecticut Ave, NW, Washington, DC 20009. 1934. 9/yr. $20; $28 to institutions.

This journal is designed to keep junior and senior high school teachers informed about new developments in science and science teaching. Approximately 300 books, 16mm films, filmstrips, and professional materials are reviewed annually by 100 science teachers across the country. The section Resources Reviews is divided into three parts: books, other books of interest, and audiovisual aids.

401. Sightlines. Nadine Covert, ed. Educational Film Library Association, 43 W 61st St, New York, NY 10023. 1967. 4/yr. free to EFLA members; $2.50 ea. to nonmembers.

The Filmlist section in each issue lists new films and videotapes alphabetically by title. Descriptive information provided by the distributor is included, but the items are not evaluated. The subject index to each Filmlist includes the heading Energy. In addition to *Sightlines,* institutional members of EFLA receive sets of film evaluation sheets 10 times a year. These cards note new films which have been critically reviewed by teams of field evaluators, including both subject specialists and media specialists. A subject index accompanies each set of cards and is cumulated annually.

402. Sources: A Guide to Print and Nonprint Materials Available from Organizations, Industry, Government Agencies, and Specialized Publishers. Maureen Crowley, ed. Gaylord Professional Publications (with Neal-Schuman Publishers), Box 61, Syracuse, NY 13201. 3/yr. $60.

Each issue describes at least 600 organizations whose publications and audiovisual productions are either not accessible through the usual bibliographic tools or who offer unusual information services to educators, librarians, and the public. A description of the particular organization's information services, such as libraries, data banks, telephone inquiry-answering, membership benefits, and reproduction services, is provided, along with a listing of representative titles of books, pamphlets, periodicals, films, slides, audiocassettes, and other nonprint materials. Each issue contains a title index; an index to free and inexpensive materials ($1.50 and under), including free-loan films; and a subject index, which is cumulative. In the 1977 index, more than 200 references to energy can be found under these particularly relevant headings: Alternative Energy Sources, Appropriate Technology, Coal and Coal Mining, Electric Power Plants, Electric Utilities, Energy, Energy Conservation, Energy Consumption, Energy Education, Energy Industry, Energy Information Systems, Energy Law, Energy Policy, Energy Research, Fuels, Gas Industry, Gas Turbines, Geothermal Energy, Heating, Insulation, Motors, Natural Gas, Northern States Power Company, Nuclear Power, Nuclear Power Plants, Petroleum Engineering, Petroleum Industry, Propane, Public Utilities, Solar Energy, Solar Heating, Solar Houses, Wind Power, and Wood Fuel.

403. Teacher. Joan Sullivan Baranski, ed. Macmillan Professional Magazines, 77 Bedford St, Stamford, CT 06901. 1882. 9/yr. $14.

Designed for teachers of grades K–8, this journal reviews approximately 500 children's books, professional books, and print and nonprint curriculum materials each year. In its "Keeping Up" department, energy-oriented materials would be noted under Science or Social Studies. Reviews, which are written by teachers, librarians, and educational consultants, indicate curriculum area and appropriate grade level.

404. Vertical File Index. H. W. Wilson Co., 950 University Ave, Bronx, NY 10452. 1935. 11/yr. w/a. cumulation. $11.

This is a subject and title index to free and inexpensive pamphlets, booklets, and mimeographed materials. A minimum of material is suitable for elementary grades and more for secondary and advanced students.

BIBLIOGRAPHIES

405. AAAS Science Book List for Children. 3rd ed. Washington, D.C.: American Association for the Advancement of Science, 1972. 253pp. $8.95; $7.95 to members.

A selected annotated list of more than 1500 science and mathematics books for children in elementary schools. Selections are based on reviews which appeared in *AAAS Science Books,* predecessor of *Science Books and Films* (see No. 399). The listing is somewhat out of date, and a new edition is scheduled for publication in fall, 1979. Fewer than five percent of the listings are for junior high level. The junior high level is the lowest level for which listings are given in the *AAAS Science Book List Supplement,* so there is little overlap between these two publications. The *Supplement,* which lists the best science books published from 1969 to 1977 and is keyed to grade level, costs $16.50, $15 to members.

406. Atomic Industrial Forum. **Audio-Visuals on Energy.** Washington, D.C.: Atomic Industrial Forum, 1977. 30pp. single copies free; $.35 ea. for 2–50 copies; $.25 ea. for over 50 copies.

This bibliography of audiovisual materials produced by industry, the government, and other sources updates previous bibliographies from AIF, the trade association for the nuclear power industry. The main emphasis is on 16mm films, but slide packages, filmstrips, and videotapes are also included. The bibliography is divided into two sections: Nuclear Subjects and General Energy Subjects. These sections are further broken down by type of producer. Critical annotations varying in length from two sentences to several paragraphs are provided for each item. Price, length, and source are also noted; audience level is not indicated for all items.

407. Balachandran, Sarojini, ed. **Energy Statistics: A Guide to Sources.** Monticello, Ill.: Council of Planning Librarians, 1976. 51pp. $5.

This is a selected, annotated guide to 162 national and international sources of statistics on all forms of energy. The contents are arranged under the following headings: Composite Sources, Coal, Electricity, Natural Gas, Nuclear Energy, Petroleum, Solar Energy, and Other. Librarians who need to check the usefulness of these statistical sources for their own clientele will find the careful annotations provide invaluable assistance. A subject index and a directory of publishers are included.

408. Bibliography of Energy Education Materials. Washington, D.C.: Center for Science in the Public Interest, Citizens Energy Project, 1977. 8pp. $.40.

This mimeographed list covers a wide range of materials from *Tilly's Catch-a-Sunbeam Coloring Book: The Story of Solar Heat Even Grownups Can Understand* to an excellent cross-section of curriculum and activity guides and manuals for teachers. Each item is thoroughly annotated, and appropriate use level is usually provided.

409. "Books on Energy: A Current Checklist." **Publishers Weekly** 212 (October 31, 1977): 33–41. New York: Bowker. $2.

This checklist covers energy titles published in or forthcoming as of 1977 by 60 publishers. Over 90 titles are listed with brief annotations under the categories: General, Oil, Alternate Sources, Nuclear, Solar, Wood, Conservation, and Architecture. Within each category, titles are arranged alphabetically by publisher. Although not a selective listing, the currency of this bibliography makes it a useful tool. The list was reprinted in *Sierra: The Sierra Club Bulletin,* 63 (February/March, 1978): 35–38.

410. Critical Mass Energy Project. **Nuclear Energy Bibliography.** Washington, D.C.: Critical Mass Energy Project, n.d. unp. free.

This bibliography is a compilation of material that has emerged in the last several years as a result of a growing citizen movement that has begun to critically examine nuclear power. It provides descriptive annotations for books, manuals, and pamphlets under the headings Background Reading and Citizen Action and is particularly useful for its references to government documents and to four Congressional committees which deal with nuclear energy. References cover materials published as recently as 1977.

411. Energy and Man's Environment. **The Energy Films Index: A Teacher's Guide to Current Energy Films.** Ed. by John Jones et al. Portland, Ore.: Energy and Man's Environment, 1977. 32pp. $4.

This index covers 183 films recommended by teachers for review by the EME staff. Each entry notes grade level, running time, and source of the film and includes a descriptive annotation. The films are not evaluated. The most recent films listed were produced in 1975; other listings date back to the 1960s. Thirty-three films are said to be available on a free-loan basis. The index is currently under revision by EME.

412. Frankena, Frederick, comp. **Energy and the Poor: An Annotated Bibliography of Social Research.** Monticello, Ill.: Council of Planning Librarians, 1977. 13pp. $1.50.

This bibliography describes 20 examples of social science research on the inter-relationship between energy consumption or pricing and different income groups in America. While most of the items would be found in a public or academic library rather than a school library, the annotations provided in the bibliography could be used as source material for student research on a topic such as the effect of rising electricity prices on the poor. Each entry is divided into method, variables, and findings, and the findings are presented with exceptional clarity.

413. Index to Environmental Studies—Multimedia. Los Angeles, Calif.: National Information Center for Educational Media, n.d. 26,000 entries. $34.50; $18.50 microfiche.

Index to Free Educational Materials—Multimedia. Los Angeles, Calif.: National Information Center for Educational Media, n.d. 10,000 entries. $26.50; $11 microfiche.

The NICEM indices are printed or microfiche catalogs which provide brief information about nonprint media available for educational use. Access is by subject or title. Each entry contains the following elements: title, subtitle, length, medium, audience level, black and white and/or color, monaural and/or stereo (for tapes and discs), Library of Congress number, date of release, producer, distributor, production code, and a brief annotation. Relevant subject headings include: Atomic Energy, Conservation, Energy, Fuel, Gas, Geothermal Energy, Nuclear Energy, Oil, Solar Energy, and Wind. Separate indices which list nonprint materials in a particular medium, e.g., *Index to 35mm Educational Filmstrips,* are also available. NICEM indices are also available online through the Lockheed DIALOG system (see Data Banks, p. 168).

414. International Solar Energy Society. **Solar Energy, Other Sources of Energy.** Killeen, Tex.: American Section of the International Solar Energy Society, 1975. unp. free w/legal-sized self-addressed, stamped envelope.

This list covers books, magazines, newsletters, reports, and plans for solar houses, heaters, and appliances which are designed for a wide variety of audiences. The emphasis is on materials concerning solar and wind energy. Some prices have not been updated since 1975, and the amount of information provided on each item varies widely. ISES states that it "has not officially approved this list of references, nor does it necessarily recommend any title herein as authoritative." A new edition is expected to be released in July, 1978.

415. Library of Congress. Congressional Research Service. **What Should Be the Energy Policy of the United States? A Preliminary Bibliography on the 1978-79 High School Debate Topic.** Washington, D.C.: Library of Congress, 1978. Dist.: Committee on Discussion and Debate. free.

This bibliography lists books, articles, and reports recommended in connection with the three resolutions suggested for the National High School Debate of 1978-79. The resolutions are: That the federal government should exclusively control the development and distribution of energy resources in the United States; That the federal government should establish a comprehensive program to significantly reduce energy consumption in the United States; That the federal government should establish a comprehensive program to significantly increase the energy independence of the United States. A final section includes general studies and articles on energy policy and some of the major sources of information on the National Energy Plan. Citations include both LC and Dewey numbers. Some items are annotated. A few sources seem too specialized for even more advanced high school students and may also be unfamiliar and inaccessible to many high school librarians.

416. Mervine, Kathryn E., and Cawley, Rebecca E. **Energy-Environment Materials Guide.** Washington, D.C.: National Science Teachers Assn., 1975. sectional paging. $2 pap.

Produced under a contract from the U.S. Office of Education's Division of Technology and Environmental Education, this comprehensive annotated bibliography is divided into four parts, each directed to a specific audience. They include: readings for teachers; readings for students grades 8–12; readings for students grades 5–9; and readings for students grades K–6. (The authors note that these usage categories are not exclusive; for example, teachers may well find references for their own use in the student bibliographies and advanced students may find items in the teacher bibliography.) Each part is further divided into 10-11 sections, ranging from general references to specific materials on electric power and energy policy. In all, about 300 items are included. Appendices include a guide to films, filmstrips, and cassettes suitable for classroom use; a list of energy-environment curriculum materials; a list of private, public, and commercial organizations which invite requests for materials; directions for obtaining government publications; and suggested methods of keeping current. An index is provided. Since this publication carries a 1975 publication date, all materials may not be available.

417. National Science Teachers Association. **Alternative Energy Sources: A Bibliography.** Oak Ridge, Tenn.: U.S. Dept. of Energy, Technical Information Center, n.d. 4pp. single copy free w/stamped, self-addressed envelope; multiple orders, $1 to cover postage.

The references included in this bibliography were selected as background documents for the preparation of the 19 National Science Teachers Association *Factsheets on Alternative Energy Technologies* (see No. 31). The bibliography covers books, pamphlets, magazine articles, and government documents. Ordering information is provided for some government publications. Items are listed under the following headings: General References, Solar Energy, Wind Power, Geothermal Energy, Energy Conservation, Nuclear Power, Coal Gasification, Energy Storage Technology, and Environmental Considerations. No annotations are provided. All items were published between 1972 and 1976.

418. National Solar Heating and Cooling Information Center. **Films/Slides: Solar Energy.** Rockville, Md.: National Solar Heating and Cooling Information Center, n.d. 14pp. free.

Regularly updated, this comprehensive annotated bibliography lists and briefly describes films, slides, and videotapes on all phases of solar energy. Sources, both commercial and government, and prices are indicated for each item. Audience level is frequently suggested, and many items have potential for student use.

419. National Solar Heating and Cooling Information Center. **Reading List for Solar Energy.** 6th ed. Rockville, Md.: National Solar Heating and Cooling Information Center, 1977. 4pp. free.

This annotated list is divided into the following sections: Nontechnical, Technical, Architectural, General Energy, Directories (refers to solar energy only), Periodicals, and Government Publications. Publishers' addresses and prices are provided; stock numbers are indicated for government publications. The brief annotations provide the only clues to the intended audience for the items listed.

420. National Solar Heating and Cooling Information Center. **Solar Energy Bibliography for Children and Young Adults.** Rockville, Md.: National Solar Heating and Cooling Information Center, 1977. lp. free.

This list of books, pamphlets, newsletters, and special publications is geared to elementary and secondary school students. The most recent item included was published in 1976. No annotations are provided, but grade levels are indicated.

421. National Solar Heating and Cooling Information Center. **Solar Energy Publications Available from the Government Printing Office.** Rockville, Md.: National Solar Heating and Cooling Information Center, n.d. lp. free.

This list notes solar publications of varying scope and audience. Because it is updated regularly, it may list materials which are not yet in the Government Printing Office's *Subject Bibliography* "Solar Energy" (see U.S. Government Sources, p. 149).

422. New York State Education Department, Bureau of Educational Communications. **The Energy Media Guide.** Albany, N.Y.: New York State Education Dept., Bureau of Educational Communications, n.d. price not available.

A list of energy films available on free loan to educators.

423. New York State Education Department, Bureau of Science Education, and Bureau of Educational Communications. **WATTS (What's Available to Teachers of Science) Bibliography.** Albany, N.Y.: New York State Education Dept., Bureau of Educational Communications, n.d. price not available.

This bibliography is keyed by energy concept and directed at teachers of science in grades K–12.

424. Outstanding Science Trade Books for Children. Reprint. New York: Children's Book Council, 1978. annual. free w/stamped, self-addressed envelope.

This annotated bibliography of the best science trade books for grades K–6 published in the preceding year is a reprint of the list published annually in a spring issue of *Science and Children* magazine. The titles are carefully selected by a book review committee appointed by the National Science

Teachers Association in cooperation with the Children's Book Council. Each review indicates appropriate grade level and is initialled by the committee member who prepared it. Although the 1978 list did not include any titles on energy, this bibliography should be considered as a potential selection aid.

425. Rhode Island Department of Education. **Energy Education Resource Guide.** Providence, R.I.: Rhode Island Dept. of Education, Dissemination Unit, n.d. 60pp. free.

This bibliographic guide is described as being "neither exclusive nor inclusive of energy education materials, nor is it an endorsement of any materials." The first four sections, General, Elementary, Secondary, and K–12, list ERIC documents and a variety of print materials available from the Dissemination Unit; each entry includes a brief annotation. Many of the items listed are curriculum programs or guides. In the last two sections, journal articles directed at professionals and resources such as books, pamphlets, periodicals, and organizations both in and outside of Rhode Island are listed.

426. Saterstrom, Mary H., comp. and ed. **Educators Guide to Free Science Materials.** 18th ed. Randolph, Wisc.: Educators Progress Service, 1977. 346pp. $11.25 pap. (19th ed. in prep., $12.75).

This multimedia bibliography is revised annually in August; the publisher advises destroying earlier editions. The 18th edition is arranged by medium: films; filmstrips, slides, and transparencies; tapes, scripts, and transcriptions; and printed materials. Films which have a restricted territorial distribution are usually not included. Each section is then divided into subject categories such as Aerospace Education, Biology, Chemistry, Environmental Education, etc. The purpose, format, length, and release date is indicated for each item, along with a brief annotation. The intended audience is not part of the standard entry, but it is sometimes included in the annotation. Titles new to this edition are starred. A source and availability index gives complete ordering information for each vendor. Advice on requesting and evaluating materials is included in the preface. Separate lists indicate which materials are available to Canadian and Australian educators. To locate energy materials, the following headings in the subject index should be consulted: Atoms and Atomic Energy, Coal and Coal Mining, Electricity, Energy and Energy Sources, Energy Crisis, Energy Storage, Nuclear Energy, Nuclear Fission, Nuclear Fusion, Petroleum and Petroleum Industry, Petroleum and Petroleum Products, and Solar Energy. The appendix contains eight study units, three of which have direct relevance to the study of energy.

427. Scherner, Sharon; Jones, John; and Dalton, Ed. **The Energy Education Bibliography: An Annotated Bibliography of Key Resources for Energy and Conservation.** 2nd. ed. Portland, Ore.: Energy and Man's Environment, 1978. 25pp. $4.

This is a current selected listing of both reference and instructional materials for energy and conservation education. More than 80 sources are given but it does not claim to be comprehensive. The publication is organized alphabetically by publisher. Preceding the annotations, the table of contents lists the material alphabetically by title and gives grade level. Space is given in the back for teachers to add other source materials.

428. U.S. Department of Energy. **Selected Department of Energy Publications.** Oak Ridge, Tenn.: U.S. Dept. of Energy Technical Information Center, 1978. 14pp. free.

This new booklet is divided into General Interest Publications and Educational Publications. There is a subject index to the former section, and many of the titles listed, e.g., *The First Reactor, Gas Heat Pumps,* and *How to Understand Your Utility Bill,* could be used in schools. The Educational Publications are primarily curriculum materials.

429. U.S. Department of Energy, Education Programs Division. **Education Publications.** Washington, D.C.: U.S. Dept. of Energy, Education Programs Division, 1978. 5pp. free.

This current multimedia bibliography is geared specifically to elementary and secondary school teachers. It is divided into two sections: materials available directly from the DOE Technical Information Center in Oak Ridge, Tennessee, and materials available from other sources. Each section is further divided into Energy Curriculum Materials and Teacher Resources. Within these areas, materials are listed by general grade categories including primary, upper elementary, junior high school, and senior high school. Precise grade levels are indicated for curriculum materials.

430. U.S. Department of the Interior. **Geothermal Energy.** Reston, Va.: U.S. Dept. of the Interior, Geological Survey, n.d. 4pp. free.

This publication lists materials on geothermal energy "chosen for their availability and their interest to general readers and students." Books, periodicals, and federal and state documents—some of which could be used by advanced secondary students—are listed.

431. U.S. Department of the Interior. **Selected References on Fossil Fuels.** Reston, Va.: U.S. Dept. of the Interior, Geological Survey, n.d. 9pp. free.

This selected listing is divided into three sections: General; Oil, Gas, Oil Shale, Asphalt: Summaries—Reserves, Production, Developments; and Coal. Materials listed span a wide range of interest and ability levels, from technical titles and government documents to popular books.

432. Weber, David O. "The Matter of Energy: A Film Perspective." **Lifelong Learning** 45 (No. 57). 1, 4-9. Berkeley, Calif.: Univ. of California Extension Media Center, 1976. free.

This informative article was published in the March 29, 1976, issue of *Lifelong Learning.* It is a bibliographic essay providing general background on energy films and describing the best of the 60 films which Weber evaluated and recommended for the Media Center's collection. (Scientists from the Lawrence Berkeley Laboratory assisted in evaluating the films.) EMC order numbers follow each title. The essay is divided into the following sections: Background Films; "Crunch" or Problem-Oriented Overviews; Alternatives in Overview; Water; Oil and Natural Gas; Coal; Nuclear-Fission Reactors; Fusion; Solar Energy; and Energy Conservation. No film produced before 1970 is included. Although audience level is not specifically indicated, the descriptions provide clues to the films' appropriate audiences.

433. Wert, Jonathan, comp. **Energy: Selected Resource Materials for Developing Energy Education/Conservation Programs.** Washington, D.C.: National Wildlife Federation, n.d. 22pp. single copy free.

This selected listing includes 62 references to resource materials developed by and available from state and regional education centers and private energy research groups and associations. Prices, annotations, and grade levels are indicated for some items.

Part Four

Further Sources of Information

Periodicals and Periodical Indices

Periodicals offer particularly current information on energy, and many general or popular science magazines now include articles on energy topics. To learn which periodicals concentrate on energy-related subjects, the librarian or educator should check *Ulrich's International Periodicals Directory, 17th Edition 1977–78* (New York: Bowker, 1977); *Magazines for Libraries*, 3rd edition, edited by Bill Katz and Berry Gargal (New York: Bowker, 1978); and *Periodicals for School Media Programs*, edited by Selma K. Richardson (Chicago: American Library Assn., 1978). However, many of the energy periodicals listed in the first two tools are not geared for use by students in elementary and secondary schools but by a technical or scientific audience.

Indices generally provide the best way to gain access to specific periodical articles or to other non-book print materials. The teacher or student should be aware of the following indices:

Applied Science and Technology Index. H. W. Wilson Co., 950 University Ave, Bronx, NY 10452. 1913. 11/yr. including q. and a. cumulations. Service basis.

Covers 297 periodicals in such relevant subject areas as energy resources and research; petroleum and gas; environment; mining; and nuclear engineering.

Business Periodicals Index. H. W. Wilson Co., 950 University Ave, Bronx, NY 10452. 1958. 11/yr. including q. and a. cumulations. Service basis.

References to articles on energy-related industries are included in this subject index to 272 periodicals.

Education Index. H. W. Wilson Co., 950 University Ave, Bronx, NY 10452. 1932. 10/yr. including q. and a. cumulations. Service basis.

This index to 330 educational periodicals, proceedings, yearbooks, and monographs covers articles on energy topics specifically treated in the curriculum and those of general interest to teachers at every level.

General Science Index. H.W. Wilson Co., 950 University Ave, Bronx, NY 10452. 1978. 10/yr. including q. and a. cumulations. Service basis.

This new Wilson index to 89 English-language science periodicals encompasses 17 science areas and is directed at the secondary, community college, and undergraduate student, as well as nonspecialist patrons in public libraries. It will cover the range of energy topics under broader headings than the *Applied Science and Technology Index* and will not make extensive use of scientific headings nor subheadings. The format parallels that of other Wilson indices. The first issue, published July, 1978, indexes periodicals beginning with issues dated June, 1978.

The New York Times Index. Microfilming Corporation of America, 21 Harristown Rd, Glen Rock, NJ 07542. a. volume, semi-m. issues, and q. cumulations, $278; semi-m. issues, $150; a. volume, $160.

This publication is an index to articles which have appeared in *The New York Times.* It is generally used in conjunction with *The New York Times on Microfilm,* which costs $650 (vesicular) or $750 (silver halide) for both the daily and Sunday papers; prices for volumes before 1960 are lower. In the *Index,* articles are listed chronologically under major headings such as Energy and Power, Atomic Energy, etc. Reference is to newspaper date, section, page, and column. Since many references provide brief details, the *Index* can be used alone as a reference tool if the microfilm is not available. Another publication, *The New York Times Subject and Personal Name Index: Environment, 1965–1975,* includes some energy events of that decade. Price is $30 to $90 depending on the institution's book and periodical budget.

Public Affairs Information Service Bulletin. Public Affairs Information Service, 11 W 40th St, New York, NY 10018. 1915. semi-m. w/q. and a. cumulations. $180 to full members; $120 to associate members (cumulations only); $85 to limited members (a. cumulation only).

PAIS uses a "broad interdisciplinary approach" to such fields as public policy, political science, economics, international relations, public administration, legislation, etc., many of which have bearing on energy topics. It indexes periodical articles, books, pamphlets, government documents, yearbooks, directories, and other English-language publications of public and private agencies throughout the world.

Readers' Guide to Periodical Literature. H. W. Wilson Co., 950 University Ave, Bronx, NY 10452. 1905. 21/yr. w/q. and a. cumulations. $55.

Abridged Readers' Guide to Periodical Literature. H. W. Wilson Co., 950 University Ave, Bronx, NY 10452. 1936. 9/yr. w/q. and a. cumulations.

These two indices are well known to almost every librarian and library user. The unabridged *Guide* now covers 179 popular, nontechnical American periodicals. The abridged version, tailored to the needs of elementary school and small public libraries, indexes 58 periodicals of general interest.

Social Sciences Index. H.W. Wilson Co., 950 University Ave, Bronx, NY 10452. 1974. q. including a. cumulation. Service basis.

This relatively new Wilson index superseded the *Social Sciences and Humanities Index* and indexes about 260 periodicals. It covers energy-related topics in such fields as economics, environmental sciences, political science, and public administration.

Subject Index to Children's Magazines. 2223 Chamberlain Ave, Madison, WI 53705. 1949. 11/yr. w/semi-a. cumulations. $12.

This index locates (by subject) materials in 61 magazines geared to elementary school, middle school, and junior high students. Some titles, such as *National Geographic* and *Popular Science,* also have a secondary audience and will be found in other indices. But the bulk are published for the elementary age and are not indexed elsewhere. Relevant subject headings include: Atomic Power, Atomic Power Plants, Automobiles (or Electric Cars); Conservation of Resources, Electric Power, Electricity, Firewood, Geothermal Energy, Heating, Hydroelectric Plants, Natural Gas, Nuclear Wastes, Petroleum, Power Crisis, Power Conservation, Power Resources, Solar Heating, Solar Power. This index may be more useful to teachers seeking materials for primary and intermediate students than to the students themselves; upper elementary students may find the *Abridged Readers' Guide* or *Readers' Guide* more useful because of the range of magazines these indices include.

With the exception of the *Times* and the *Subject Index to Children's Magazines,* these indices use many of the same subject headings. In addition, each uses headings relevant to its own particular focus. The following is a list of subject headings which may be useful when doing research. No single index utilizes them all. To allow for students doing research on past energy events, references to United States government agencies which are no longer in existence are included.

Air Conditioning Equipment	Electric Utilities—Costs
Atomic Power	Electric Utilities—Environmental Aspects
Atomic Power Plants	Electric Utilities—Export-Import Problems
Automobiles—Energy Conservation Measures	Electric Utilities—Fuel Requirements
Automobiles—Energy Use	Electric Utilities—Rates
Biomass Energy	Energy Conservation
Buildings—Energy Usage	Energy Industries
Coal	Energy Policy
Coal Gasification	Energy Policy (divided geographically)
Coal Industry	Energy Resources
Coal Leases	Fuel
Coal Pipe Lines	Fuel, Synthetic
Economic Conditions	Fuel Economy
Electric Power	Fuel Supply
Electric Power Plants	Gas, Natural
Electric Utilities—Atomic Power Activities	Gasoline Industry
Electric Utilities—Competitive Fuel	Hydroelectric Power

Liquefied Petroleum Gas
Magnetohydrodynamics
Nuclear Fission
Nuclear Fuels
Nuclear Fusion
Nuclear Reactions—Fission
Nuclear Reactions—Fusion
Nuclear Reactors
Nuclear Reactors—Fuel Reprocessing
Oil and Gas Leases
Oil Fields
Oil Pollution of Rivers, Harbors etc.
Oil Pollution of the Sea
Oil Reclamation
Oil Refineries
Oil Sands
Oil Shale Industry
Oil Shales
Oil Well Drilling
Oil Well Drilling, Submarine
Oil Well Drilling Rigs
Oil Wells, Submarine
Organization of Petroleum Exporting
   Countries
Petroleum
Petroleum, Synthetic
Petroleum in Submerged Lands
Petroleum Industry
Petroleum Industry—Safety Measures
Petroleum Laws and Regulations
Petroleum Pipelines
Petroleum Refineries
Petroleum Supply
Petroleum Waste
Plutonium
Power Resources (divided geographically)
Power Resources—Conservation

Power Resources—Economic Aspects
Power Resources—Environmental Aspects
Power Resources—Statistics
Power Resources—Supply and Demand
Radioactive Substances
Radioactive Waste Disposal
School Buildings—Energy Conservation
   Measures
Solar Batteries
Solar Energy
Solar Energy Industry
Solar Engines
Solar Furnaces
Solar Heat Collectors
Solar Heating
Solar Housing
Solar Power
Solar Research
Steam, Geothermal
Steam, Natural
Tank Ships
Tidal Power
Tide Power
United States—Atomic Energy Commission
United States—Economic Conditions
United States—Energy, Department of
United States—Energy Policy
United States—Energy Research and
   Development Administration
United States—Federal Energy Administration
United States—Nuclear Regulatory
   Commission
Uranium
Water Heaters
Wind Power
Windmills

Energy companies are also excellent sources of energy periodicals. For example, many oil companies publish magazines as well as stockholders' reports. For an extensive list of American oil companies, see Industrial, Citizen, Professional, and Public Sources, p. 165. Some of these companies will send publications only to libraries, not to individual teachers. Their publications are frequently found in the reference departments of public or university libraries or the business sections of large public libraries.

U.S. Government Sources

The federal government is probably the largest single publisher of energy materials. The main tools for locating federal publications are the following guides. These guides can be ordered from the Superintendent of Documents, U.S. Government Printing Office, Washington, D.C. 20402.

Subject Bibliographies (free) list current sale publications on specific topics available from the Government Printing Office. Most energy materials will be listed in the bibliographies entitled "Atomic Energy and Nuclear Power" (No. 200), "Energy Conservation and Resources" (No. 58), and "Solar Energy" (No. 9).

Monthly Catalog of U.S. Government Publications (12/yr. w/annual indexes, $45) is the most complete listing of government publications. It includes both Congressional documents and Departmental publications issued during the previous month. Entries are arranged by issuing agency. Author, title, subject, and series report indices are provided in each issue and cumulated semi-annually and annually. Subscription includes a serials supplement. The subject index now uses Library of Congress subject headings.

Selected U.S. Government Publications (10/yr. free) lists approximately 150 to 200 popular government publications available for sale from the Government Printing Office. Relevant materials will be found in the section "Energy and Environment" included in each issue.

The newly created Department of Energy, established in October, 1977, is now the primary government source of materials and information about energy. The DOE was established by the Department of Energy Organization Act (approved on August 4, 1977), which consolidated all major federal energy functions into one Cabinet-level Department. All of the responsibilities of the Energy Research and Development Administration, the Federal Energy Administration, the Federal Power Commission, and the Alaska, Bonneville, Southeastern and Southwestern Power Administrations, formerly components of the Department of the Interior, as well as the power marketing functions of the Department of the Interior's Bureau of Reclamation, were transferred to the Department of Energy. DOE also assumed responsibility for certain functions of the Interstate Commerce Commission, the Department of Commerce, the Department of Housing and Urban Development, the Department of Defense–Navy, and the Department of the Interior. Many books, pamphlets, and bibliographies still refer to publications of the agencies or parts of agencies now a part of the Department of Energy, but queries directed to these superseded agencies will probably be routed to DOE.

The Department of Energy now offers a variety of brochures, pamphlets, charts, and curriculum materials (see Selection Aids, Nos. 428, 429). An updated film catalog, *28 Energy Films,* of free-loan films available from the Department, will be published in the summer of 1978. It will list gen-

Energy: A Multimedia Guide for Children and Young Adults
JUDITH H. HIGGINS

ERRATA

Here is the complete text for missing page 150:

eral-interest films, arranged by subject. All DOE materials, including this catalog, should be requested on institutional letterhead from the U.S. Department of Energy, Technical Information Center, Box 62, Oak Ridge, TN 37830. Further inquiries about these films, beyond what is included in the catalog, should be addressed to the Department of Energy, Office of Public Affairs, Audiovisual Branch, Washington, DC 20545. Those who are interested in purchasing the films, rather than acquiring them on loan, should request the catalog that is also entitled *28 Energy Films* from the General Services Administration, National Archives and Records Service, The National Audiovisual Center, Order Section F, Washington, DC 20409.

A complete *Energy Films Catalog* listing 105 free-loan films keyed by subject and audience level will be available by September, 1978. It should also be ordered from the Department of Energy, Technical Information Center.

Another source of information on energy films is the Department of the Interior's *1976 Film Catalog: Bicentennial, Environmental and Natural Resources Films*. Films concerning energy are listed under the categories "Water and Power Resources"; "Electric Power—Pacific Northwest, Pacific Southwest"; "Mineral Resources"; and "States and Their Natural Resources." It is available free from the U.S. Department of the Interior, Washington, DC 20240.

The following federal government agencies may be contacted directly for other energy-related materials which may be appropriate for use with students in elementary and secondary schools. (The National Solar Heating and Cooling Information Center is a quasi-governmental agency operated by the Franklin Institute Research Laboratories for the Department of Energy and the Department of Housing and Urban Development.)

General Services Administration
Office of Information
18th and F Sts. NW. Washington, DC 20405

National Aeronautics and Space
  Administration
Lewis Research Center, Room 056A, 600
  Independence Ave, NW, Washington, DC
  20546

National Solar Heating and Cooling
  Information Center
Box 1607, Rockville, MD 20850

U.S. Department of Commerce
National Bureau of Standards
Administration Bldg, Room A-607,
  Washington, DC 20234

U.S. Department of Health, Education and
  Welfare
Office of Consumer Affairs
330 Independence Ave. SW, Washington,
  DC 20201

U.S. Department of Labor
Community Services Administration
1200 19th St, NW, Washington, DC 20212

U.S. Department of State
Bureau of Public Affairs
2201 C St, NW, Washington, DC 20520

U.S. Department of the Interior
Bureau of Mines, Office of Mineral
  Information
Columbia Plaza, 2401 E St, NW,
  Washington, DC 20241

U.S. Department of the Interior
Bureau of Reclamation, Publications
  Branch
18th and C Sts, NW, Room 7444,
  Washington, DC 20240

U.S. Department of the Interior
Geological Survey, Geologic Inquires
  Group
Mail Stop 907, Reston, VA 22092

Free-loan and Rental Films

Films which are available at no cost or for a low rental fee may convey technical information better than print material can, and for this reason they are recommended to schools and libraries introducing energy topics to classes or audiences with relatively little technical background.

Many educators are already accustomed to dealing, either directly or indirectly, with industrial firms which supply these films. The caveat in the Introduction regarding books and other print materials applies doubly here: because information is presented in 16mm or videotape form does not mean it necessarily presents a fair picture of the topic. Unfortunately, some teachers take a less critical view of films; they provide no introduction or follow-up and let ride unchallenged certain biases which they would not allow to pass in print. Users of the free-loan films should be aware that typically these films are "sponsored," that is, their cost of production and distribution is underwritten by a company or industry. They should be alert for the subtle or not-so-subtle message of the sponsor and be prepared to explain and discuss this message with students. The list of free-loan film sources which follows is arranged alphabetically by organization name.

Free-loan Films and Videotapes

ASSOCIATION FILMS
866 Third Ave, New York, NY 10022

Association distributes free-loan films made available by industries, associations, and foundations to community organizations as an information service. Their catalog, *Association Films: Free-Loan Films,* lists many energy-connected films under the heading "Conservation, Energy, Environment" in the topical index; however, the entire catalog should be checked. Association maintains ten film exchanges in the United States and three in Canada.

EXXON U.S.A.
Box 2180, Houston, TX 77001

Seven of the 11 free-loan films listed in the brochure "The Exxon Film Library" are directly related to energy topics. They may be ordered from 19 Exxon film libraries around the United States.

GIFT PROGRAM
New York State Education Department, Bureau of Educational Communications
Albany, NY 12234

The GIFT (Government and Industrial Films for Teaching) Program offers a selection of video programs on energy to New York State teachers. The GIFT Energy Package contains 26 video programs derived from sponsored films and covering most areas of energy. By December, 1978, each program will have a rewritten annotation emphasizing key energy-instruction concepts and the program's strong points and biases, plus additional materials such as discussion questions. Some programs contain pre- and post-tests as well.

New York State public, private, and parochial schools may obtain these programs through their regional Board of Cooperative Educational Services educational communications center. Teachers in Buffalo, New York City, Rochester, and Syracuse should contact their supervisors of educational communications. Small cities and districts not participating in other BOCES programs, and private and parochial schools, may elect to participate in the GIFT program through arrangement with their nearest BOCES. The BOCES may reproduce and make available GIFT programs in any video format.

MODERN TALKING PICTURE SERVICE
2323 New Hyde Park Rd, New Hyde Park, NY 11040

Modern Talking Picture Service distributes free-loan films from 24 film libraries in the U.S. and three in Canada. The bulk of its energy-related films are listed in catalogs called *Free-Loan Films for Elementary Schools* and *Science and Technology: Free-Loan Educational Films,* directed at secondary teachers. Individual titles may be requested. MTPS recommends, however, that teachers take part in its Preferred Program Service, through which titles in certain categories—some of which are new and do not appear in their brochures—are shipped on a list of requested dates or on a regular schedule. To receive energy films through this service, the categories to check are the subheadings "Petroleum and Related Products," "Mineral Extraction and Processing," and "Ecology," under the major heading "Science." MTPS recommends that orders for any titles be placed as early in the school year as possible, preferably the preceding spring.

MODERN VIDEO CENTER
2323 New Hyde Park Rd, New Hyde Park, NY 11040

Modern Video Center, a leading distributor of free-loan videocassette programs, offers seven full-color television programs on energy topics. These programs vary in length from 15 to 28 minutes and are compatible with U-matic ¾ inch videocassette systems.

SCREEN NEWS DIGEST
235 E 45th St, New York, NY 10017

Four energy films are available on a free-loan basis from this firm. Screen News Digest provides a discussion outline for the teacher with each film. The outline includes persons, places, and facts to watch for in the film, questions and answers, and suggestions for related classroom activities.

SHELL FILM LIBRARY
1433 Sadlier Circle West Dr, Indianapolis, IN 46239

Four of the 25 films listed in the library's current free-loan film list relate directly to energy: *The Fossil Story, Refinery Processes, Oil Well,* and *Oil,* but others touch on related subjects such as pollution.

SOHIO FILM LIBRARY
2515 Franklin Blvd, Cleveland, OH 44113

Sohio films are primarily for schools and groups meeting in Ohio. Energy-connected titles include three on the construction of the Alaskan pipeline; *Refining,* which explains the basic principles of refining crude oil to a non-technical audience; and *Energy in Perspective,* which dicusses man's historic use of energy, the limits of fossil fuels, and alternative energy sources.

University Film Rental Collections

In addition to direct rental from producers or distributors, teachers or librarians may rent energy films from college and university film-rental collections. Only film libraries which do not limit rentals to particular groups and which have a total collection of more than 100 titles (not necessarily on energy) are included in the list of university film rental collections provided here. The list is arranged alphabetically by state.

University of Arizona
Bureau of Audiovisual Services
1325 East Speedway, Tucson, AZ 85719

University of California, Berkeley
Extension Media Center
Berkeley, CA 94720

University of Southern California
Division of Cinema, Film Distribution
  Center
University Park, Los Angeles, CA 90007

University of Colorado
Educational Media Center, Film Library
Stadium Bldg, Boulder, CO 80309

University of Delaware
Instructional Resources Center
Newark, DE 19711

Rentals are limited to Delaware.

Florida State University
Regional Film Library
Tallahassee, FL 32306

University of Idaho
AV/Photo Center
Moscow, ID

Rentals are limited to Idaho, Oregon, and Washington.

Northern Illinois University
Media Distribution Department
Altgeld Hall 114, DeKalb, IL 60115

Rentals are limited to the continental U.S.

Southern Illinois University
Film Rental Library, Learning Resources
Service
Carbondale, IL 62901

University of Illinois
Visual Aids Service
1325 S Oak St, Champaign, IL 61820

Rentals are limited to the continental U.S.

Western Illinois University
Media Center
Macomb, IL 61455

Rentals are limited to West Central Illinois.

Indiana State University
Audio-Visual Center
Terre Haute, IN 47809

Prefer to rent in Midwest only.

Indiana University
Audio-Visual Center
Bloomington, IN 47401

Iowa State University
Media Resources Center
121 Pearson Hall, Ames, IA 50001

Rentals are limited to the continental U.S.

University of Iowa
Media Library - Audiovisual Center
C-5 East Hall, Iowa City, IA 52242

Rentals are limited to continental U.S.

University of Kansas
Film Rental Services, Audio-
Visual Center
746 Massachusetts St, Lawrence, KS 66044

University of Kentucky
Audio-Visual Services
Scott St Bldg, Lexington, KY 40506

Rentals are limited to 48 states.

Boston University
School of Education, Krasker
Memorial Film Library
765 Commonwealth Ave, Boston, MA 02215

Rentals are limited to 48 states.

Michigan State University
Instructional Media Center, Off-Campus
Scheduling Office
East Lansing, MI 48824

University of Michigan
Audio-Visual Education Center
416 Fourth St, Ann Arbor, MI 48109

Wayne State University
Film Library, Systems Distribution and
Utilization Department
Detroit, MI 48202

University of Minnesota
Audio Visual Library Service
3300 University Ave E, Minneapolis,
MN 55414

Southeast Missouri State University
Eastern Missouri Film Company
Audiovisual Center
900 Normal Ave, Cape Girardeau, MO 63701

Rentals are limited to Missouri
and surrounding states.

University of Missouri—Columbia
Academic Support Center Film Library
505 E Stewart Rd, Columbia, MO 65201

Rentals are limited to continental U.S.

University of Montana
Instructional Materials Service and
U.S.D.A. Forest Service,
Northern Region Film Library
Missoula, MT 59812

Rentals are limited to Montana and
areas of adjacent states.

University of New Hampshire
Department of Media Services
Durham, NH 03824

Rentals are limited to New England states.

Cornell University
Film Library, Media Services
Box 47, Roberts Hall, Ithaca, NY 14853

Rentals are limited to 48 states.

New York University
Film Library
26 Washington Place, New York, NY 10003

State University College at Buffalo
Film Rental Library—Communication
Center
1300 Elmwood Ave, Buffalo, NY 14222

Rentals are limited to 50 states.

State University of New York at Buffalo
Educational Communication Center,
Media Library
1 Foster Annex, Buffalo, NY 14214

Rentals are limited to 48 states.

Syracuse University
Film Rental Center
1455 East Colvin St, Syracuse, NY 13210

Rentals are limited to major Northeast
and Southeast states.

North Dakota State Film Library
Division of Independent Study
State University Sta, Fargo, ND 58102

Rentals are limited to North Dakota,
South Dakota, and Minnesota.

Kent State University
Audio Visual Services
Kent, Ohio 44242

Rentals are limited to 48 states.

Oregon State System of Higher Education
Continuing Education Film Library
1633 SW Park Ave, Box 1491, Portland,
OR 97207

Rentals are limited to 13 Western states.

The Pennsylvania State University
Audio Visual Services
Special Services Bldg,
University Park, PA 16802

Rentals are limited to continental U.S.

University of South Carolina
Instructional Services Center
Columbia, SC 29208

South Dakota State University
Audio Visual Center
Pugsley 101, Brookings, SD 57007

The University of Texas at Austin
Film Library
Drawer W, University Sta, Austin, TX 78712

Rentals are limited to continental U.S.

Brigham Young University
Audiovisual Services
290 HRCB, Provo, UT 84602

University of Utah
Educational Media Center
207 Milton Bennion Hall,
Salt Lake City, UT 84112

Rentals are limited to 48 states.

Utah State University
Audiovisual Services
UMC 31, Logan, UT 84322

Rentals are limited to 48 states.

Shoreline Community College
16101 Greenwood Ave, N, Seattle,
WA 98133

Rentals are limited to the state
of Washington.

Washington University
Instructional Media Services
Pullman, WA 99163

Rental charge is double outside of
Washington, Oregon, Idaho, and Montana.

University of Wisconsin—Extension
Bureau of Audio-Visual Instruction
1327 University Ave, Box 2093,
Madison, WI 53701

There is a $10 surcharge for films
sent outside of Wisconsin.

University of Wisconsin—LaCrosse
Film Library
1705 State St, LaCrosse, WI 54601

University of Wyoming
Film Library—Audio-Visual Services
Box 3273, University Sta,
Laramie, WY 82071

Rentals are limited to 50 states.

Industrial, Citizen, Professional, and Public Sources

In addition to the types of sources listed elsewhere in this guide, there are firms, associations, agencies, and special-interest organizations—some with strong biases—which publish energy materials. While this material may not have been specifically designed for students, much of it can be useful to them. Teachers and librarians should also consider sources such as local or regional utility companies and citizens' groups, which may make material available only to individuals or organizations in their own areas. Following are an annotated list of sources of various materials on energy and a nonannotated list of oil companies known to have either magazines or annual stockholders' reports, or both.

THE ALUMINUM ASSOCIATION
818 Connecticut Ave, NW, Washington, DC 20006

A trade organization representing over 81 companies, including all the primary producers of aluminum in the United States, this association offers several free-loan films and filmstrips and single copies (free) of reprinted articles connecting energy with aluminum production and use. Other publications include a free eight-page "Teachers' Resource Guide," which is a directory of free and inexpensive classroom materials, and a booklet, "Energy and the Aluminum Industry" ($2), which describes and charts recommendations of this energy-intensive industry regarding a national energy policy and the industry's part in the economy and in energy conservation.

AMERICAN ASSOCIATION FOR THE ADVANCEMENT OF SCIENCE
1776 Massachusetts Ave, NW, Washington, DC 20036

The largest general scientific organization representing all fields of science, the AAAS has many committees concerned with aspects of energy. The *AAAS Science Film Catalog,* published in 1975, is still available from the R.R. Bowker Co. for $16.95.

AMERICAN PETROLEUM INSTITUTE
2101 L St, NW, Washington, DC 20036

This national trade association encompasses all branches of the petroleum industry. Its many free publications include "Gasoline for Your Car"; a 44-page picture booklet, "Facts about Oil"; "Diesel Fuel Questions and Answers"; a series of brief pamphlets entitled "Alternative Energy Sources";

a catalog of general interest films called "Movies about Oil"; an annually-updated catalog called "Publications and Materials"; and an up-to-date source book, *Looking for Energy: A Guide to Information Resources,* in which many items are keyed by grade level.

AMERICAN SECTION OF THE INTERNATIONAL SOLAR ENERGY SOCIETY
Box 1416, U.S. Hwy 190W, Killeen, TX 76541

This group seeks to foster the science and application of solar energy and to encourage solar energy basic and applied research, development, and demonstration. It provides a list of a range of solar-related books and their publishers. Student memberships to ISES or the American Section are available for $12 each or $21 for a combined membership.

BREEDER REACTOR CORPORATION
Box U, Oak Ridge, TN 37830

The Breeder Reactor Corporation is a group of 750 electric systems around the country who have sponsored the Clinch River Breeder Reactor Plant Project, a cooperative effort of the federal government and the utility industry to build the first large-scale demonstration breeder nuclear power plant. A number of publications explaining the breeder reactor are available free from the Corporation. They include a booklet, "Questions and Answers about the Liquid Metal Fast Breeder Reactor," and pamphlets such as "Facts and Figures about the Clinch River Breeder Reactor Plant Project"; "Capsule Summary: Why We Need the Breeder"; and "How Much Radiation Will the Public Receive from the Clinch River Breeder Reactor Plant?"

CALIFORNIA STATE ENERGY RESOURCES CONSERVATION AND DEVELOPMENT COMMISSION
1111 Howe Ave, Sacramento, CA 95825

Educators and librarians may request and fill out an application giving preferences among the categories of material they want from this agency. Included are materials on appliance standards, energy legislation, thermal power, and social impacts of energy use and development. Other free materials, sent anywhere in the world, include the "Solar Information Packet 1977"; a booklet, "How to Conquer the Energy Peak"; a cartoon booklet, "Saving Energy at Home, It's Your Money"; and a poster, *Help Later Gator Conquer the Energy Peak.*

CITIBANK PETROLEUM DEPARTMENT
399 Park Ave, New York, NY 10022

This department publishes the quarterly *Energy Newsletter,* covering financial and economic aspects of the energy industries. It is free of charge if requested in writing.

CITIZENS' ENERGY PROJECT
Center for Science in the Public Interest, 8th floor, 1413 K St, NW, Washington, DC 20005

A nonprofit environmental and consumer research organization which publishes and distributes information to encourage citizen involvement in energy decision-making. Publications include *Nuclear Power and Civil Liberties; Can We Have Both?* ($4 to individuals, $7 to organizations); *People and Energy* magazine ($10 to individuals, $16 to libraries); and the *Citizens' Energy Directory* ($7.50), which lists 500 individuals and organizations working on alternative energy systems, as well as leading energy magazines and newsletters.

COMMITTEE FOR NUCLEAR RESPONSIBILITY
Main Box 11207, San Francisco, CA 94101

A nonprofit organization dedicated to public education about the problems of nuclear power and the advantages of other energy systems. Single copies of CNR publications are free and may be reproduced. The current publications list includes short, introductory flyers such as "Nuclear Power . . . Bad for Health and Life . . . Bad for the Economy"; Lewis Mumford's 1974 address on solar energy, "Enough Energy for Life"; a three-part analysis called "Jimmy Carter's Energy Plan: Myths vs. Realities"; and "Alice in Blunderland," Dr. John Gofman's one-hour presentation in his nuclear power debate against Dr. Edward Teller.

CONSOLIDATED EDISON COMPANY OF NEW YORK
4 Irving Place, New York, NY 10003

This New York public utility provides a "Catalogue of Available Programs, Exhibits and Literature from Con Edison" to schools and to professional, social, or community groups in its service area. People who wish to borrow films without a Con Edison speaker should telephone their nearest Con Edison Consumer Affairs office. Teaching materials include a kit, *Gas Conservation,* for grades 4–9; comic books such as "Where the Little Light Bulb Gets Its Juice"; an "Energy Management Checklist" to encourage conservation at home; and a series of "Energy Saver" conservation pamphlets.

CRITICAL MASS ENERGY PROJECT
Box 1538, Washington, DC 20013

More than 175 local citizens' groups now work with CMEP on a wide range of issues involving nuclear power, energy conservation, and the development of renewable energy technologies. CMEP publishes the monthly *Critical Mass Journal* which informs the general public on energy issues as they relate to regulatory agencies, legislative matters, and citizen activism. It also issues legislative updates and "action alerts" to cooperating local groups to apprise them of important federal energy developments and to encourage greater public participation in energy policy decision-making. Among CMEP's other publications are citizen's manuals such as *Nuclear Plants: The More They Build the More You Pay* ($5 to individuals) and *Citizens' Guide to Nuclear Power* ($5).

EDUCATIONAL FACILITIES LABORATORIES
850 Third Ave, New York, NY 10022

EFL is a nonprofit organization which researches and provides information on the building and operation of facilities for public educational institutions. One service is called Energy and Schools: Public Information Program; it is funded by the Exxon Corporation to provide materials to concerned citizens about energy conservation in schools. One publication of this program is the brochure "Questions and Answers on Energy Conservation in Schools." For this brochure and additional information regarding the program, contact EFL. A second program called Public Schools Energy Conservation Service uses computer-based models to analyze school buildings and to report on operating and capital improvements which will save energy. For enrollment information, write John Boice, EFL, 3000 San Hill Rd, Bldg 1, Suite 120, Menlo Park, CA 94025. In 1978 EFL published a revised and updated edition of *The Economy of Energy Conservation in Educational Facilities* (96 pp., $4). This title contains strategies for making energy conservation programs mandatory on all remodeled and new construction projects, for identifying and correcting sources of waste in existing buildings, and for selecting equipment with low long-range operating costs. This 1978 edition also contains information on solar-generated heating for both new and existing buildings.

ENVIRONMENTAL ACTION
1346 Connecticut Ave, NW, Suite 731, Washington, DC 20036

An outgrowth of Earth Day, Environmental Action engages in research and public education. Its energy-related publications include *How to Challenge Your Local Electric Utility* ($2.50). A newsletter, *Environmental Activist,* explains the state of energy measures in Congress and what citizens can do to assist. It and a bi-weekly magazine, *Environmental Action,* are sent to members. Single copies of the magazine are available for $.75.

ENVIRONMENTAL ACTION COALITION
156 Fifth Ave, Suite 1130, New York, NY 10010

This nonprofit organization seeks to identify specific environmental problems that affect New York City and other urban areas across the nation. Seven times yearly, it publishes *Eco-News: An Environmental Newsletter for Young People* ($6.95 with a 40% discount on multiple copies, grades 5-9), which highlights energy as well as environmental issues. Its relevant curriculum unit is *Less Power to the People: Environmental Energy Use,* a packet containing lesson plans, issues of *Eco-News,* teacher guides on energy, bibliographies, and suggested audiovisual materials. The packet is $5.20 to members, $6.50 to non-members, and is suggested for use with grades 5 and up.

ENVIRONMENTAL INFORMATION CENTER OF THE FLORIDA
    CONSERVATION FOUNDATION
935 Orange Ave, Winter Park, FL 32789

This center performs research and provides information on environmental issues of importance to Florida. It is the source for the Summer, 1976, issue ($5) of the *Florida Scientist,* which contains a long article on practical applications of solar energy in Florda, as well as other solar research papers of a technical nature.

EXXON COMPANY U.S.A.
Exxon Publications Library, Public Affairs Department, Box 2180,
    Houston, TX 70001

This major oil company offers free-loan films (see p. 151), and a number of free items listed in its *Exxon Publications Library for College Reference* can be used by younger students. These range from comic books, such as "Mickey Mouse and Goofy Explore Energy" and "Mickey Mouse and Goofy Explore Energy Conservation," to a four-color chartbook, "Energy Outlook 1977–1990," and a technical overview, "Technology in the Search for Energy." Useful publications from a "background series" include "World Energy Outlook," "Middle East Oil," and "The Offshore Search for Oil and Gas." Some of the "oil industry issues" series can be used by advanced secondary students; titles include "OPEC: Questions and Answers," "The Case for Diversification," "Energy and Economic Progress," and "The Charges and the Facts," a discussion of the fallacies they see underlying the suggestion that the vertically integrated oil companies be broken up and prohibited from diversifying into non-petroleum energy.

NATIONAL PETROLEUM COUNCIL
1625 K St, NW, Washington, DC 20006

The Council advises, informs, and makes recommendations to the Secretary of the Interior on any matter related to petroleum or the petroleum industry. It publishes in-depth reports such as *Potential for Energy Conservation in the United States: 1979-1985* ($8.50); *Ocean Petroleum Resources* ($6.95); *Emergency Preparedness for Interruption of Petroleum Imports into the United States* (summary report, $1.50); and a series of reports on the U.S. energy outlook.

NATURAL RESOURCES DEFENSE COUNCIL
122 E 42nd St, New York, NY 10017

A nonprofit organization, dedicated to protecting America's natural and human resources, the Council publishes a quarterly *NRDC Newsletter* ($15) which includes energy topics. A recent issue covered the war over the expansion of electric power in the Pacific Northwest and conservation and efficiency advantages of solar energy.

PACIFIC GAS AND ELECTRIC COMPANY
Public Information Department, 77 Beale St, San Francisco, CA 94108

PG&E operates the largest geothermal electric-generating installation in the world—The Geysers, located some 90 miles northeast of San Francisco. It is also the only commercial geothermal electric installation in the United States. By 1979 its cumulative plant capacity will be 908,000 kilowatts. The company offers free an illustrated brochure, "The Geysers," and other information describing its development of geothermal energy. A film, *The Ballad of Steamy Valley*, is available only to groups within its service area. Visits to The Geysers may be scheduled by groups or individuals.

POWER AUTHORITY OF THE STATE OF NEW YORK
10 Columbus Circle, New York, NY 10019

The authority finances, builds, and operates electric-generating facilities and transmission lines for purposes specified by the legislature and governor of New York. Its free booklet, "Electric Energy and the Environment," describes the operation of the Authority's nuclear and fossil-fueled plants and hydroelectric facilities.

PUBLIC CITIZEN HEALTH RESEARCH GROUP
Dept P, 2000 P St, NW, Washington, DC 20036

A Ralph Nader group which publishes information on atomic industry workers and consumer goods using radioactive material.

RECORDING FOR THE BLIND
215 E 58 St, New York, NY 10022

About a dozen energy-science titles are among the nearly 40,000 master tapes available on open-reel or cassette to visually, physically, or perceptually handicapped persons who cannot read ordinary material. Elementary and high school students may request that additional titles be recorded in the period December through July; complete instructions regarding this process should be requested from RFB. The current catalog, with supplements, is available for $3.

RESOURCES FOR THE FUTURE
1755 Massachusetts Ave, NW, Washington, DC 20036

RFF is a private, nonprofit organization for research and education to advance the development, conservation, and use of natural resources. It publishes *Resources* (free) three times yearly. In the Fall, 1976, issue of *Resources,* Marc K. Landy presents a clearly written primer on the ecology and economics of Kentucky strip mining, excerpted from a book also published by RFF.

SEATTLE CITY LIGHT
Community Relations Department, City Light Bldg, 1015 Third Ave, Seattle, WA 98104

This electric utility distributes energy-related coloring charts for the primary grades, a sheet on how to read your electric meter, an energy-wise checklist, films for both elementary and secondary levels, and a number of energy conservation booklets suitable for consumer education or vocational education classes. These publications are available free to customers, teachers, and students within their service area. In addition, the company offers locally-oriented pamphlets and tours of three different electrical facilities, one of which is a solar-wind experiment.

SHELL OIL COMPANY
Public Affairs, One Shell Plaza, Box 2463, Houston, TX 77001

Shell distributes free a variety of educational materials, including the factual picture booklet "Shell's Wonderful World of Oil" (Gr. 6–12) and longer booklets such as the 45-page "The Story of Petroleum." *Shell Views, Shell Reports,* and position papers, all reflecting the oil industry viewpoint on energy topics such as industry dismemberment, environmental conservation, and the National Energy Act, are available for more sophisticated secondary readers.

SOCIETY OF PETROLEUM ENGINEERS OF AIME
6200 N Central Expwy, Dallas, TX 75206

This professional group offers a 20-page booklet, "Careers in Petroleum Engineering" (free) and a nontechnical kit, *Oil from the Earth* (79 color slides w/1 cassette and 1 booklet, $40), which illustrates the application of engineering and science to the discovery, development, and production of petroleum. Loan copies of the kit are also available.

TENNESSEE ENERGY OFFICE
250 Capitol Hill Bldg, Nashville, TN 37219

Coal is Tennessee's most abundant natural fuel resource. A 127-page publication from this state office, *Coal in Tennessee,* answers questions about the state's coal production, use, distribution, and research.

TEXACO
Public Relations Department, 2000 Westchester Ave, White Plains, NY 10650

Among the free educational materials available from this energy company are "Protecting the Environment," which describes and illustrates the company's environmental activities, and a scriptographic booklet, "What Everyone Should Know about Offshore Energy."

UNION OF CONCERNED SCIENTISTS
1208 Massachusetts Ave, Cambridge, MA 02138

This independent nonprofit watchdog is led by professionals who feel that nuclear power is a high-cost energy source with serious, unresolved safety-problems. They advocate "prudent energy planning" emphasizing solar energy over nuclear power. Among their free and low cost pamphlets are "Answers to Your Questions about Nuclear Dangers . . . and Solar Potential," "Facts about the Nuclear Power Industry," and "Do You Know What Plutonium Is?"

UNIPUB
Box 433, Murray Hill Sta, New York, NY 10016

UNIPUB is the U.S. distributor of publications of the United Nations system and other international information publishers. Although most of the publications listed are technical, the catalog *International Publications on Energy Resources* also lists materials such as *Power Reactors in Member States* ($8), *Nuclear Power: The Fifth Horseman* ($2), *Energy: The Case for Conservation* ($2), and *Energy Crisis and the Future* ($4).

ZOMEWORKS CORPORATION
Box 712, Albuquerque, NM 87103

This group specializes in consultation, design, and research on passive solar systems. The publications list includes *Solar Water Heating Plans* ($5), *Drum Wall Plans* ($5), and a set of plans and specifications for a Beadwall system of insulation ($15).

Oil Companies

Amerada Hess Corporation
1185 Ave of the Americas, New York, NY 10019

American Petrofina
Box 2159, Dallas, TX 75221

Aminoil U.S.A.
Box 94193, Houston, TX 77018

Amoco Oil Company
200 E Randolph Dr, Chicago, IL 60601

Ashland Oil
1401 Winchester Ave, Ashland, KY 41101

Atlantic Richfield Company
515 S Flower St, Los Angeles, CA 90071

BP Oil
3515 Silverside Rd, Clayton Bldg, Wilmington, DE 19810

Sub. of Standard (Ohio)

BP North America
620 Fifth Ave, New York, NY 10020

Champlin Petroleum Company
5301 Camp Bowie Blvd, Box 9365, Fort Worth, TX 76107

Cities Service Company
Cities Service Bldg, Box 300, Tulsa, OK 74102

Clark Oil and Refining Corporation
8530 W National Ave, Milwaukee, WI 53227

Continental Oil Company
Box 2197, Houston, TX 77001

Crown Central Petroleum Corporation
One N Charles, Box 1168, Baltimore, MD 21203

Diamond Shamrock Oil and Gas Company
Box 631, Amarillo, TX 79173

Exxon Company U.S.A.
Box 2180, Houston, TX 77001

Getty Oil Company
3810 Wilshire Blvd, Los Angeles, CA 90010

Gulf Oil Corporation
Gulf Bldg, Pittsburgh, PA 15230

Hunt Oil Company
First National Bank Bldg, 1401 Elm, Dallas, TX 75202

Husky Oil Company
Box 280, Cody, WY 82414

Kerr-McGee Corporation
Kerr-McGee Center, Oklahoma City, OK 73102

Marathon Oil Company
539 S Main St, Findlay, OH 45840

Mobil Oil Corporation
150 E 42nd St, New York, NY 10017

Murphy Oil Corporation
200 Jefferson Ave, El Dorado, AR
71730

Pennzoil Company
700 Milam, Houston, TX 77002

Phillips Petroleum Company
Phillips Bldg, Bartlesville, OK 74004

Quaker State Oil Refining Corporation
11 Center St, Oil City, PA 16301

Shell Oil Company
One Shell Plaza, Box 2463, Houston, TX
77001

Skelly Oil Company
                    See Getty Oil Company

Standard Oil Company of California
225 Bush St, San Francisco, CA 94104

Standard Oil Company (Indiana)
200 E Randolph Dr, Chicago, IL 60601

The Standard Oil Company (Ohio)
Midland Bldg, Cleveland, OH 44115

Sun Company
100 Matsonford Rd, Radnor, PA 19087

Tenneco Oil Company
Box 2511, Houston, TX 77001

Texaco
135 E 42nd St, New York, NY 10017

Union Oil Company of California
Union Oil Center, Los Angeles, CA 90017

Data Bases

Data bases are included for the benefit of teachers and librarians who are doing research for their own collections or who wish to direct very advanced secondary students to a wider range of sources. In each case, the students are advised to check their nearest college or university library regarding the availability and cost of computer searches. Three advantages to data base searches are: 1) search requests may be tailored to specific needs, 2) the speed of recall in locating items is more rapid and offers more confrontations than painstaking searches of a number of printed volumes covering the same period, and 3) there are more subject access points to the data base because, for most data bases, there is what is sometimes called full-text searching capability, i.e., an article or record can be retrieved through any word (except commonly used prepositions and articles) in the title or in an abstract. Some pertinent data bases follow:

ENVIRONMENT INFORMATION CENTER
292 Madison Ave, New York, NY 10017

EIC is a private clearinghouse for energy and environment information which offers data base services, books, and abstracts which are available on a subscription basis. Those data bases relating to energy are:

Energyline. An online data base that provides access to abstracts from periodicals, hard-to-find non-serials, and important mass media ranging from *Energy Policy* to *The Wall Street Journal,* scientific/technical sources, and socioeconomic materials. The data base was established in January, 1976, and contains environment-related material beginning with 1971. It may be searched by a controlled vocabulary of approximately 5000 key terms or by the major subject categories: U.S. Economics, U.S. Policy and Planning, International, Research and Development, General, Resources and Reserves, Petroleum and Natural Gas Resources, Coal Resources, Unconventional Resources, Solar Energy, Fuel Processing, Fuel Transport and Storage, Electric Power Generation, Electric Power Storage and Transmission, Nuclear Resources and Power, Thermonuclear Power, Consumption and Conservation, Industrial Consumption, Transportation Consumption, Residential Consumption, Environmental Impact. Energyline is now available on SDC/ORBIT as File Energy and on Lockheed/DIALOG as File 69.

Enviroline. Similar to Energyline, this data base covers a wide range of journal literature, conference proceedings, speeches, government documents, and hard-to-find non-serials and irregular serials in abstract form. The major categories which are relevant to energy are Air Pollution, Energy, Land Use and Misuse, Oceans and Estuaries, and Water Pollution.

THE INFORMATION BANK (THE NEW YORK TIMES)
One World Trade Center, Suite 86011, New York, NY 10048

This data base has almost 1.4 million items online. The bank consists of virtually all the news and editorial matter from the final late city edition of *The New York Times* (including Sunday feature sections and daily and Sunday regional material not distributed within New York City) dating back to January 1, 1969. It also contains news and editorial material from 60 other publications back to January 1, 1973. Unless a student or teacher has access to an Information Bank terminal through a large organization, a library, or a company which is a subscriber, the only access is by an "on-demand" search at a cost of $50 plus $1 per retrieved item. The searcher receives an abstract of the item by mail. Full text microfiche copy from *The New York Times* is also available. These searches can be arranged through one of the Bank's six sales offices in New York and Canada.

INFORMATION CENTER FOR ENERGY SAFETY
Oak Ridge National Laboratory
Box Y, Bldg 9764, Oak Ridge, TN 37830

The Information Center for Energy Safety was established as a national center for collecting, storing, evaluating, and disseminating safety information related to the development and use of several non-nuclear forms of energy. The data bases which are available for search are largely technical in nature but provide excellent source, background, and historical review material useful for all levels of research. The following subject categories are searchable: Solar; Coal; Coal Conversion and Utilization; Oil, Gas, and Shale Technology; Magnetohydrodynamics; Thermonuclear; Geothermal; Wind; Electrical Energy Systems; Transportation and Storage; and Advanced Systems. State-of-the-art reviews of the safety of various energy systems, periodic selected dissemination of information (SDI) reports, and answers to technical inquiries are available. Unfunded students and others doing unsponsored research may receive free service.

LOCKHEED INFORMATION SYSTEMS
Code 5020/201
3251 Hanover St, Palo Alto, CA 94304

Lockheed provides online searching of more than 75 databases through its *DIALOG* information retrieval service. DIALOG offers access to Energyline, Enviroline, ERIC, NICEM and PAIS (see Periodicals and Periodical Indexes, p. 146), and to Magazine Index, a new data base created for general reference librarians which provides coverage of more than 350 popular American magazines. The cost for searching depends upon the online connect-hour rate per data base, the amount of time spent online, the number of citations printed offline and any additional telecommunications charges incurred. DIALOG indicates that a 10–12 minute search would cost between $5 and $15.

SYSTEM DEVELOPMENT CORPORATION
2500 Colorado Ave, Santa Monica, CA 90406

SDC Search Service offers online searching of a number of scientific, technical, social science, and business data bases, including Energyline. Charges are by the computer connect-hour plus a per citation cost for offline printing and a telecommunications charge. SDC recommends that students use the service of university libraries where computer literature searching is an established practice.

Directory of Publishers and Distributors

This directory provides ordering addresses for all numbered entries. Addresses of additional publishers and distributors of energy materials appear in the Further Sources of Information section.

ABINGDON PRESS
201 Eighth Ave S, Nashville, TN 37202

ABT PUBLICATIONS
55 Wheeler St, Cambridge, MA 02138

ACADEMY FILMS DISTRIBUTION
COMPANY
Box 3414, Orange, CA 92665

ADDISON-WESLEY PUBLISHING
COMPANY
Jacob Way, Reading, MA 01867

AGENCY FOR INSTRUCTIONAL
TELEVISION
Box A, Bloomington, IN 47401

ALLYN AND BACON,
470 Atlantic Ave., Boston, MA 02210

ALTERNATIVE SOURCES OF ENERGY
Route 2, Box 90A, Milaca, MN 56353

THE ALUMINUM ASSOCIATION
Communications Department, 818
Connecticut Ave, NW, Washington, DC
20006

AMERICAN ASSOCIATION FOR THE
ADVANCEMENT OF SCIENCE
1776 Massachusetts Ave, NW, Washington,
DC 20036

AMERICAN GAS ASSOCIATION
Educational Services, 1515 Wilson Blvd,
Arlington, VA 22209

AMERICAN LIBRARY ASSOCIATION
50 E Huron St, Chicago, IL 60611

THE AMERICAN MUSEUM OF
ATOMIC ENERGY/OAK RIDGE
ASSOCIATED UNIVERSITIES
Box 117, Oak Ridge, TN 37830

AMERICAN PETROLEUM INSTITUTE
Public Relations Department, 2101 L St,
NW, Washington, DC 20037

AMERICAN PUBLIC POWER
ASSOCIATION
Publications Department, 2600 Virginia
Ave, NW, Washington, DC 20037

AMERICAN SECTION OF THE
INTERNATIONAL SOLAR ENERGY
SOCIETY
Box 1416, U.S. Highway 190W, Killeen,
TX 76541

AMERICAN UNIVERSITIES FIELD
STAFF
4 W Wheelock St, Box 150, Hanover, NH
03755

ANCHOR PRESS/DOUBLEDAY
245 Park Ave, New York, NY 10017

APPALSHOP FILMS
Box 743, Whitesburg, KY 41858

ASSOCIATED EDUCATIONAL
MATERIALS COMPANIES
Box 2087, Raleigh, NC 27602

ASSOCIATION FILMS
866 Third Ave, New York, NY 10022

ATLANTIC MONTHLY PRESS/LITTLE,
BROWN AND COMPANY
34 Beacon St, Boston, MA 02114

ATOMIC INDUSTRIAL FORUM
Public Affairs and Information Program,
7101 Wisconsin Ave, NW, Washington
DC 20014

BFA EDUCATIONAL MEDIA
Box 1795, 2211 Michigan Ave, Santa
Monica, CA 90406

BALLANTINE BOOKS
201 E 50th St, New York, NY 10022

BALLINGER PUBLISHING COMPANY
17 Dunster St, Cambridge, MA 02138

BANTAM BOOKS
666 Fifth Ave, New York, NY 10019

BARR FILMS
Box 5667, 3490 E Foothill Blvd, Pasadena,
CA 91107

BENCHMARK FILMS
145 Scarborough Rd, Briarcliff Manor,
NY 10510

CHANNING L. BETE
45 Federal St, Greenfield, MA 01301

STEPHEN BOSUSTOW PRODUCTIONS
1649 11th St, Santa Monica, CA 90404

R. R. BOWKER COMPANY
1180 Ave of the Americas, New York,
NY 10036
or
Magazine Subscriptions, Box 67, Whitinsville, MA 01588

BOY SCOUTS OF AMERICA
Supply Division, North Brunswick, NJ
08902

BREEDER REACTOR CORPORATION
Box U, Oak Ridge, TN 37830

BULLFROG FILMS
Oley, PA 19547

CAVALCADE PRODUCTIONS
Box 801, Wheaton, IL 60187

CENTER FOR INFORMATION ON
AMERICA
Washington, CT 06793

CENTER FOR SCIENCE IN THE
PUBLIC INTEREST
1413 K St, NW, 8th Floor, Washington
DC 20005

CENTRON EDUCATIONAL FILMS
1621 W Ninth St, Lawrence, KS 66044

CHANGING TIMES EDUCATION
SERVICE
1729 H St, NW, Washington, DC 20006

CHESHIRE BOOKS
Church Hill, Harrisville, NH 03450

CHILDREN'S BOOK COUNCIL
67 Irving Place, New York, NY 10003

CHILDREN'S SCIENCE BOOK REVIEW
COMMITTEE
Longfellow Hall, 13 Appian Way,
Cambridge, MA 02138

CHURCHILL FILMS
662 N Robertson Blvd, Los Angeles,
CA 90069

CITATION PRESS
50 W 44th St, New York, NY 10036

CITIBANK
Petroleum Department, 399 Park Ave,
New York, NY 10022

CLARION/SEABURY SERVICE
CENTER
Customer Service, Somers, CT 06071

CLEARVUE INC.
6666 N Oliphant Ave, Chicago, IL 60631

COLUMBIA UNIVERSITY PRESS
562 W 113th St, New York, NY 10025

COMMITTEE FOR ECONOMIC
DEVELOPMENT
477 Madison Ave, New York, NY 10022

COMMITTEE FOR NUCLEAR
RESPONSIBILITY
Main PO Box 11207, San Francisco, CA
94101

COMMITTEE ON DISCUSSION AND
DEBATE
224 Jessup Hall, Iowa City, IA 52242

CONCEPT MEDIA
1500 Adams Ave, Costa Mesa, CA 92626

CONCERN INC.
2233 Wisconsin Ave, NW, Washington, DC
20007

CONGRESSIONAL QUARTERLY
1414 22nd St, NW, Washington, DC 20037

CONSOLIDATED EDISON COMPANY
OF NEW YORK
4 Irving Place, New York, NY 10003

CONSOLIDATION COAL COMPANY
Public Relations Department, One Oliver
Plaza, Pittsburgh, PA 15222

CORONET FILMS
65 E South Water St, Chicago, IL 60601

CORONET INSTRUCTIONAL MEDIA
65 E South Water St, Chicago, IL 60601

COUNCIL OF PLANNING
LIBRARIANS
Box 229, Monticello, IL 61856

COUNCIL ON ECONOMIC PRIORITIES
84 Fifth Ave, New York, NY 10011

COWARD, McCANN AND
GEOGHEGAN
200 Madison Ave, New York, NY 10016

CRITICAL MASS ENERGY PROJECT
Box 1538, Washington, DC 20013

THOMAS Y. CROWELL COMPANY
10 E 53rd St, New York, NY 10022

CRYSTAL PRODUCTIONS
Airport Business Center, Box 11480,
Aspen, CO 81611

CURRENT AFFAIRS
24 Danbury Rd, Wilton, CT 06897

JOHN DAY COMPANY
10 E 53rd St, New York, NY 10022

DILLON PRESS
500 S Third St, Minneapolis, MN 55415

WALT DISNEY EDUCATIONAL
MEDIA COMPANY
400 S Buena Vista St, Burbank, CA 91521

DODD, MEAD AND COMPANY
79 Madison Ave, New York, NY 10016

DOMUS BOOKS
c/o Quality Books, 400 Anthony Trail,
Northbrook, IL 60062

DONARS PRODUCTIONS
Educational Materials, Box 24, Loveland,
CO 80537

DOUBLEDAY AND COMPANY
245 Park Ave, New York, NY 10017

DOVER PUBLICATIONS
180 Varick St, New York, NY 10014

DUTCHESS COUNTY BOARD OF
COOPERATIVE EDUCATIONAL
SERVICES
RD 1, Salt Point Turnpike, Poughkeepsie,
NY 12601

ERIC CLEARINGHOUSE ON
EDUCATIONAL MANAGEMENT
University of Oregon, Eugene, OR 97403

ERIC/SMEAC INFORMATION
REFERENCE CENTER
The Ohio State University, College of
Education, 1200 Chambers Rd, 3rd Floor,
Columbus, OH 43212

EARTHMIND/PEACE PRESS
5246 Boyer Rd, Mariposa, CA 95338

EDISON ELECTRIC INSTITUTE
Public Relations Department, 90 Park
Ave, New York, NY 10016

THOMAS ALVA EDISON
FOUNDATION
Suite 143, Cambridge Office Plaza,
18280 W Ten Mile Rd, Southfield, MI
48075

EDITORIAL RESEARCH REPORTS
1414 22nd St, NW, Washington, DC 20037

EDUCATIONAL ACTIVITIES
Freeport, NY 11520

EDUCATIONAL DESIGN
47 W 13th St, New York, NY 10011

EDUCATIONAL DIMENSIONS GROUP
Box 126, Stamford, CT 06904

EDUCATIONAL FILM LIBRARY
ASSOCIATION
43 W 61st St, New York NY 10023

EDUCATIONAL IMAGES
Box 367, Lyons Falls, NY 13368

EDUCATIONAL INSIGHTS
20435 S Tillman Ave, Carson, CA 90746

EDUCATIONAL MATERIALS AND
EQUIPMENT COMPANY
Box 17, Pelham, NY 10803

EDUCATIONAL RESEARCH COUNCIL
OF AMERICA
Rockefeller Bldg, 614 W Superior, Cleveland,
OH 44113

EDUCATORS PROGRESS SERVICE
214 Center St, Randolph, WI 53956

ELECTRICAL INDUSTRIES
ASSOCIATION OF SOUTHERN
CALIFORNIA
Suite 630, 9911 W Pico Blvd, Los Angeles,
CA 90035

ENCORE VISUAL EDUCATION
1235 S Victory Blvd, Burbank, CA 91502

ENCYCLOPAEDIA BRITANNICA
EDUCATIONAL CORPORATION
425 N Michigan Ave, Chicago, IL 60611

ENERGY AND MAN'S ENVIRONMENT
0224 SW Hamilton, Suite 301, Portland,
OR 97201

ENERGY CONSERVATION
RESEARCH
9 Birch Rd, Malvern, PA 19355

ENERTECH
Box 420, Norwich, VT 05055

ENVIRONMENT INFORMATION
CENTER
292 Madison Ave, New York, NY 10017

ENVIRONMENTAL ACTION
1346 Connecticut Ave, NW, Suite 731,
Washington, DC 20036

ENVIRONMENTAL ACTION
COALITION
156 Fifth Ave, Suite 1130, New York, NY
10010

ENVIRONMENTAL ACTION
FOUNDATION
1346 Connecticut Ave, NW, Suite 731,
Washington, DC 20036

ENVIRONMENTAL ACTION REPRINT
SERVICE (EARS)
2239 E Colfax, Denver, CO 80206

ENVIRONMENTAL INFORMATION
CENTER OF THE FLORIDA
CONSERVATION FOUNDATION
935 Orange Ave, Winter Park, FL 32789

ENVIRONMENTALISTS FOR FULL
EMPLOYMENT
1101 Vermont Ave, NW, Room 305,
Washington, DC 20005

M. EVANS AND COMPANY
216 E 49th St, New York, NY 10017

EXXON PUBLICATIONS LIBRARY
Public Affairs Department, Exxon
Company U.S.A., Box 2180, Houston, TX
77001

EYE GATE MEDIA
146-01 Archer Ave, Jamaica, NY 11435

FACTS ON FILE
119 W 57th St, New York, NY 10019

FARRAR, STRAUS AND GIROUX
19 Union Sq W, New York, NY 10003

FIELD MUSEUM OF NATURAL
HISTORY
Roosevelt Rd at Lake Shore Dr, Chicago,
IL 60605

STUART FINLEY
3428 Mansfield Rd, Falls Church, VA 22041

FORD FOUNDATION
320 E 43rd St, New York, NY 10017

FOREIGN POLICY ASSOCIATION
345 E 46th St, New York, NY 10017

FOUR WINDS PRESS
50 W 44th St, New York, NY 10036

THE FREE PRESS
866 Third Ave, New York, NY 10022

W. H. FREEMAN AND COMPANY
660 Market St, San Francisco, CA 94104

FRIENDS OF THE EARTH
529 Commercial St, San Francisco,
CA 94111

GAF CORPORATION
Audio/Visual Order Department,
Binghamton, NY 13902

GARDEN WAY PUBLISHING
Charlotte, VT 05445

GARRARD PUBLISHING COMPANY
1607 N Market St, Champaign, IL 61820

GAYLORD PROFESSIONAL
PUBLICATIONS
Box 61, Syracuse, NY 13201

GENERAL SERVICES
ADMINISTRATION
Office of Information, 18th and F Sts, NW,
Washington, DC 20405

GREEN MOUNTAIN POST FILMS
Box 177, Montague, MA 01351

STEPHEN GREENE PRESS
Box 1000, Brattleboro, VT 05301

THE GRIST MILL
90 Depot Rd, Eliot, ME 03903

GROSSMAN PUBLISHERS
625 Madison Ave, New York, NY 10022

GRUMMAN AEROSPACE
CORPORATION, HORIZONS
MAGAZINE
Dept. 930-05, Bethpage, NY 11714

GUIDANCE ASSOCIATES
757 Third Ave, New York, NY 10017

GULF OIL CORPORATION
Consumer Affairs EEE, Box 1563,
Houston, TX 77001

HANDEL FILM CORPORATION
8730 Sunset Blvd, West Hollywood, CA
90069

HARCOURT BRACE JOVANOVICH
757 Third Ave, New York NY 10017

HARPER AND ROW PUBLISHERS
10 E 53rd St, New York, NY 10022

HARRIS COUNTY DEPARTMENT
OF EDUCATION
Instructional Center, 6208 Irvington Blvd,
Houston, TX 77022

HARVEY HOUSE PUBLISHERS
20 Waterside Plaza, New York, NY 10010

HASTINGS HOUSE PUBLISHERS
10 E 40th St, New York, NY 10016

HAWKHILL ASSOCIATES
125 E Gilman, Madison, WI 53703

HIGHLINE PUBLIC SCHOOLS,
PROJECT ECOLOGY
15675 Ambaum Blvd, SW, Seattle, WA 98166

HORN BOOK
Park Square Bldg, 31 St. James Ave,
Boston, MA 02116

INDEPENDENT PETROLEUM
ASSOCIATION OF AMERICA
1101 16th St, NW, Washington, DC 20036

INDIANA UNIVERSITY
AUDIO-VISUAL CENTER
Bloomington, IN 47401

INDIANA UNIVERSITY PRESS
Tenth and Morton Sts, Bloomington, IN
47401

THE INFORMATION BANK (NEW
YORK TIMES)
Suite 86011, One World Trade Center,
New York, NY 10048

INSTITUTE FOR LOCAL
SELF-RELIANCE
1717 18th St, NW, Washington DC 20009

INSTITUTE OF GAS TECHNOLOGY
Document Center, 3424 S State St,
Chicago, ILL 60616
Chicago, IL 60616

THE INSTRUCTOR PUBLICATIONS
Instructor Park, Dansville, NY 14437

IOWA ENERGY POLICY COUNCIL
710 E Locust, Des Moines, IA 50309

THE JOHNS HOPKINS UNIVERSITY
PRESS
Baltimore, Maryland 21218

KENT STATE UNIVERSITY AUDIO
VISUAL SERVICES
Kent, OH 44242

THE KIRKUS SERVICE
200 Park Ave S, New York, NY 10003

WALTER J. KLEIN LTD.
6301 Carmel Rd, Charlotte, NC 28211

ALFRED A. KNOPF
201 E 50th St, New York, NY 10022

LANDERS ASSOCIATES
Box 69760, Los Angeles, CA 90069

LEARNING SYSTEMS COMPANY
1818 Ridge Rd, Homewood, IL 60430

LEE COUNTY SCHOOLS
ENVIRONMENTAL EDUCATION
PROGRAM
2055 Central Ave, Fort Myers, FL 33901

LITTLE, BROWN AND COMPANY
34 Beacon St, Boston, MA 02114

LORIEN HOUSE
BOX 1112, Black Mountain, NC 28711

M. I. T. PRESS
28 Carleton St, Cambridge, MA 02142

MACMILLAN LIBRARY SERVICE
866 Third Ave, New York, NY 10022

MACMILLAN PROFESSIONAL
MAGAZINES
77 Bedford St, Stamford, CT 06901

MACMILLAN PUBLISHING COMPANY
866 Third Ave, New York, NY 10022

McGRAW-HILL BOOK COMPANY
1221 Ave of the Americas, New York,
NY 10020

McGRAW-HILL FILMS
1221 Ave of the Americas, New York,
NY 10020

DAVID McKAY COMPANY
750 Third Ave, New York, NY 10017

MATSON'S
Box 308, Milltown, MT 59851

MENTOR BOOKS/NEW AMERICAN
LIBRARY
1301 Ave of the Americas, New York,
NY 10019

CHARLES E. MERRILL PUBLISHING
COMPANY
1300 Alum Creek Dr, Columbus, OH 43216

JULIAN MESSNER
1230 Ave of the Americas, New York,
NY 10020

MICHIGAN ASSOCIATION OF
SCHOOL ADMINISTRATORS
421 W Kalamazoo, Lansing, MI 48933

MODERN TALKING PICTURE
SERVICE
2323 New Hyde Park Rd, New Hyde
Park, NY 11040

MODERN VIDEO CENTER
2323 New Hyde Park Rd, New Hyde
Park, NY 11040

ARTHUR MOKIN PRODUCTIONS
17 W 60th St, New York, NY 10023

MORGAN AND MORGAN
145 Palisades St, Dobbs Ferry, NY 10522

WILLIAM MORROW AND COMPANY
105 Madison Ave, New York, NY 10016

MULTI-MEDIA PRODUCTIONS
Box 5097, Stanford, CA 94305

NATIONAL COAL ASSOCIATION
Education Division, Coal Bldg, 1130
17th St, NW, Washington, DC 20036

NATIONAL ELECTRIC RELIABILITY
COUNCIL
Research Park, Terhune Rd, Princeton,
NJ 08540

NATIONAL FILM BOARD
OF CANADA
1251 Ave of the Americas, New York,
NY 10020
or
Box 6100, Montreal, Quebec, Canada
H3C 3H5

NATIONAL GEOGRAPHIC SOCIETY
Box 2806, Washington, DC 20013

NATIONAL INFORMATION CENTER
FOR EDUCATIONAL MEDIA
(NICEM)
University of Southern California,
University Park, Los Angeles, CA 90007

NATIONAL PETROLEUM COUNCIL
1625 K St, NW, Washington, DC 20006

NATIONAL SCIENCE TEACHERS
ASSOCIATION
1742 Connecticut Ave, NW, Washington,
DC 20009

NATIONAL SOLAR HEATING AND
COOLING INFORMATION CENTER
Box 1607, Rockville, MD 20850

THE NATIONAL WILDLIFE
FEDERATION
1412 16th St, NW, Washington, DC 20036

NATURAL RESOURCES DEFENSE
COUNCIL
122 E 42nd St, New York NY 10017

NEBRASKA EDUCATIONAL
TELEVISION COUNCIL FOR
HIGHER EDUCATION
Box 83111, Lincoln, NE 68501

NEW AMERICAN LIBRARY
1301 Ave of the Americas, New York,
NY 10019

THE NEW FILM COMPANY
331 Newbury St, Boston MA 02115

NEW JERSEY EDUCATION
ASSOCIATION
180 W State St, Trenton, NJ 08608

NEW YORK STATE COLLEGE OF
AGRICULTURE AND LIFE
SCIENCES, CORNELL UNIVERSITY
Riley-Robb Hall, Ithaca, NY 14853

NEW YORK STATE COLLEGE OF
HUMAN ECOLOGY AND NEW
YORK STATE COLLEGE OF AGRI-
CULTURE AND LIFE SCIENCES
c/o Mailing Room, 7 Research Park, Cornell
University, Ithaca, NY 14853

NEW YORK STATE EDUCATION
DEPARTMENT
Room 314G, Albany, NY 12234

NEW YORK STATE EDUCATION
DEPARTMENT, BUREAU OF
EDUCATIONAL   COMMUNICATIONS
Albany, NY 12234

NEWSBANK INC.
Box 10047, 741 Main St, Stamford, CT
06904

NORTH DAKOTA STATE FILM
LIBRARY
Division of Independent Study, State
University Sta, Fargo, ND 58102

NORTHERN ILLINOIS GAS COMPANY
Attn: Film Librarian, Route 59 and EW
Tollway, Naperville, IL 60540

W. W. NORTON AND COMPANY
500 Fifth Ave, New York, NY 10036

OAK RIDGE ASSOCIATED
UNIVERSITIES, ENERGY
EDUCATION DIVISION
Box 117, Oak Ridge, TN 37830

OCEANA PUBLICATIONS
75 Main St, Dobbs Ferry, NY 10522

ORES PUBLICATIONS
University of Kentucky, College of
Engineering, Lexington, KY 40508

PACIFIC GAS AND ELECTRIC
COMPANY
Public Information Department, 77 Beale
St, San Francisco, CA 94108

PENGUIN BOOKS
625 Madison Ave, New York, NY 10022

PENNSYLVANIA DEPARTMENT OF
EDUCATION, OFFICE OF
INFORMATION AND PUBLICATIONS
Box 911, Harrisburg, PA 17126

PERSPECTIVE FILMS
369 W Erie St, Chicago, ILL 60610

PILOT BOOKS
347 Fifth Ave, New York, NY 10016

POWER AUTHORITY OF THE STATE
OF NEW YORK
10 Columbus Circle, New York, NY 10019

PRENTICE-HALL
Englewood Cliffs, NJ 07632

PRENTICE-HALL MEDIA
150 White Plains Rd, Tarrytown, NY
10591

PROFESSIONAL ARTS
Box 8003, Stanford, CA 94305

PUBLIC AFFAIRS CLEARINGHOUSE
Box 30, Claremont, CA 91711

PUBLIC CITIZEN HEALTH RESEARCH
GROUP
Department P, 2000 P St, NW, Washington,
DC 20036

PUBLIC INTEREST RESEARCH
GROUP
Box 19312, Washington, DC 20036

PUBLIC ISSUES NETWORK
1755 Massachusetts Ave, NW, Washington,
DC 20036

PUBLIC TELEVISION LIBRARY
Video Program Service, 475 L'Enfant
Plaza, SW, Washington, DC 20024

Q-ED PRODUCTIONS
Box 1608, 2921 W Alameda Ave, Burbank,
CA 91507

RMI MEDIA PRODUCTIONS
701 Westport Rd, Kansas City, MO 64111

RAINTREE CHILDREN'S BOOKS
205 W Highland Ave, Milwaukee, WI 53203

RAMSGATE FILMS
704 Santa Monica Blvd, Santa Monica,
CA 90401

RANDOM HOUSE EDUCATIONAL
MEDIA
400 Hahn Rd, Westminster, MD 21157

REAL GAS AND ELECTRIC COMPANY
Box F, Santa Rosa, CA 95402

RECORDING FOR THE BLIND
215 E 58th St, New York, NY 10022

RESOURCES FOR THE FUTURE
1755 Massachusetts Ave, NW, Washington,
DC 20036

RHODE ISLAND DEPARTMENT OF
EDUCATION, DISSEMINATION
UNIT
22 Hayes St, Providence, RI 02908

WARD RITCHIE PRESS
474 S Arroyo Pkwy, Pasadena, CA 91105

RODALE PRESS
Emmaus, PA 18049

RICHARDS ROSEN PRESS
29 E 21st St, New York, NY 10010

SCHOLAR'S CHOICE LTD.
50 Ballantyne Ave, Stratford, Ontario,
Canada N5A 6T9

SCHOLASTIC BOOK SERVICES
50 W 44th St, New York, NY 10036

SCHOLASTIC MAGAZINES
50 W 44th St, New York, NY 10036

SCIENCE AND MANKIND
2 Holland Ave, White Plains, NY 10603

SCIENTISTS' INSTITUTE FOR
PUBLIC INFORMATION
355 Lexington Ave, New York, NY 10017

SCREEN NEWS DIGEST
235 E 45th St, New York, NY 10017

SHAWNEE MISSION PUBLIC SCHOOLS
Environmental Education and Science
Resource Center, Marsha Bagby School,
95th and Mohawk, Shawnee Mission,
KS 66206

SHEED ANDREWS AND McMEEL
6700 Squibb Rd, Mission, KS 66202

SIERRA CLUB BOOKS
530 Bush St, San Francisco, CA 94108

SIERRA CLUB INFORMATION
SERVICES
530 Bush St, San Francisco, CA 94108

BILL SNYDER FILMS
Box 2784, Fargo, ND 58102

SOCIAL ISSUES RESOURCES SERIES
1200 Quince Orchard Blvd, Gaithersburg,
MD 20760

SOCIETY FOR VISUAL EDUCATION
1345 Diversey Pkwy, Chicago, IL 60614

SOCIETY OF PETROLEUM
ENGINEERS OF AIME
6200 N Central Expwy, Dallas, TX 75206

SPARTAN GRAPHICS
288 S State St, Sparta, MI 49345

STACKPOLE BOOKS
Cameron and Keller Sts, Harrisburg, PA
17105

STANDARD PROJECTOR AND
EQUIPMENT COMPANY
3080 Lake Terrace, Glenview, IL 60025

STECK-VAUGHN COMPANY
Box 2028, Austin, TX 78768

STRUCTURES PUBLISHING COMPANY
Box 423, Farmington, MI 48024

SUNBEAM APPLIANCE COMPANY
Consumer Affairs Department, 2001 S York
Rd, Oak Brook, IL 60521

SYSTEM DEVELOPMENT
CORPORATION
2500 Colorado Ave, Santa Monica, CA
90406

TAB BOOKS
Blue Ridge Summit, PA 17214

TENNESSEE ENERGY AUTHORITY
Suite 250, Capitol Hill Bldg, Nashville,
TN 37219

TIME-LIFE BOOKS
c/o Silver Burdett Company, 250 James
St, Morristown, NJ 07960

TIME-LIFE MULTIMEDIA
100 Eisenhower Dr, Paramus, NJ 07652

TIME-WISE PUBLISHERS
Box 4140, Pasadena, CA 91106

UNITED NATIONS ENVIRONMENT
PROGRAMME
New York Liaison Office, United Nations,
New York, NY 10017

U. S. DEPARTMENT OF ENERGY,
   EDUCATION PROGRAMS DIVISION
Washington, DC 20545

U.S. DEPARTMENT OF ENERGY,
   TECHNICAL INFORMATION
   CENTER
Box 62, Oak Ridge, TN 37830

U.S. DEPARTMENT OF THE
   INTERIOR, GEOLOGICAL SURVEY
Geologic Inquiries Group, Mail Stop 907,
   Reston, VA 22092

U.S. GOVERNMENT PRINTING
   OFFICE
Washington, DC 20402

UNIVERSITY OF CALIFORNIA
   EXTENSION MEDIA CENTER
Berkeley, CA 94720

UNIVERSITY OF CHICAGO
   GRADUATE LIBRARY SCHOOL,
   UNIVERSITY OF CHICAGO PRESS
1100 E 57th St, Chicago, IL 60637

UNIVERSITY OF COLORADO,
   EDUCATIONAL MEDIA CENTER
Stadium Bldg, Boulder, CO 80309

UNIVERSITY OF MASSACHUSETTS
   PRESS
Box 429, Amherst, MA 01002

UNIVERSITY OF WISCONSIN SEA
   GRANT COLLEGE PROGRAM
1800 University Ave, Madison, WI 53706

VERMONT CROSSROADS PRESS
Box 333, Waitsfield, VT 05673

VIKING PRESS
625 Madison Ave, New York, NY 10022

VINTAGE/RANDOM HOUSE
201 E 50th St, New York, NY 10022

VIRGINIA ENERGY OFFICE
823 E Main St, Richmond, VA 23219

VISUAL EDUCATION CONSULTANTS
Box 52, Madison, WI 53701

VISUAL EDUCATION COPORATION
Box 2321, Princeton, NJ 08540

VISUAL PURPLE
Box 996, Berkeley, CA 94701

WALKER AND COMPANY
720 Fifth Ave, New York, NY 10019

WARD'S NATURAL SCIENCE
   ESTABLISHMENT
Box 1712, Rochester, NY 14603
   or
Box 1749, Monterey, CA 93940

FRANKLIN WATTS
730 Fifth Ave, New York, NY 10019

WESTMINSTER PRESS
Witherspoon Bldg Room 905, Philadelphia,
   PA 19107

H. W. WILSON COMPANY
950 University Ave, Bronx, NY 10452

WORLD BOOK-CHILDCRAFT
   INTERNATIONAL
Educational Services, Sta 8, Merchandise
   Mart, Chicago, IL 60654

WORLDWATCH INSTITUTE
1776 Massachusetts Ave, NW, Washington,
   DC 20036

XEROX EDUCATION PUBLICATIONS
1250 Fairwood Ave, Box 444, Columbus,
   OH 43216

XEROX FILMS
245 Long Hill Rd, Middletown, CT 06457

YALE UNIVERSITY PRESS
302 Temple St, New Haven, CT 06520

ZOMEWORKS CORPORATION
Box 712, Albuquerque, NM 87103

Author Index

Title Index

Subject Index